SCIENCE OF
CHAOS

S C I E N C E
O F
CHAOS

CHRISTOPHER
LAMPTON

A Venture Book
Franklin Watts
New York / London / Toronto / Sydney

Photographs courtesy of the author except the following photographs
copyright ©: National Oceanic and Atmospheric Administration/National
Weather Service: p. 15; North Carolina Supercomputing Center: p. 22
(simulation by Charles Evans at the University of North Carolina/Chapel
Hill, visualization by Ray Idaszak); Los Alamos National Laboratory/Fred
Rick: p. 57; Historical Pictures Service, Chicago: p. 62; Massachusetts
Institute of Technology/Jane McNabb: p. 66; NASA: p. 81; *Scientific
American*: p. 86 (from SA, Feb. 90, "Chaos and Fractals in Human
Physiology" by Ary L. Goldberger, David R. Rigney, Bruce Smith).

Library of Congress Catalog-in-Publication Data
Lampton, Christopher.
Science of Chaos / by Christopher Lampton.
p cm. —(A Venture book)
Includes bibliographical references and index.
Summary: A discussion of a new scientific theory formed
with the help of computers, which proposed the possibility
of patterns and order in the chaos and unpredictability of
both nature and human aspects of the world.
ISBN 0-531-12513-0
1. Chaotic behavior in systems—Juvenile literature.
[1. Chaotic behavior in systems.] I.Title.
Q172.5.c45L36   1992
003'.7—D20   91-40896   CIP   AC

Contents

Chapter One
A World in Chaos
9

Chapter Two
Simple Numbers, Simple World
17

Chapter Three
Numbers in the Fast Lane
26

Chapter Four
Simple Numbers, Complicated World
37

Chapter Five
Chaos in the Air
59

Chapter Six
Chaos Everywhere
72

Chapter Seven
Pictures of Chaos
84

Epilogue
109

Appendix
Typing BASIC Programs
111

Source Notes
115

Glossary
117

For Further Reading
121

Index
123

SCIENCE OF
CHAOS

C H A P T E R

ONE

A World in Chaos

The screeching drone of a jet aircraft cuts across the sky, followed by the deafening burst of a sonic boom.

A sleek black plane flashes above a desert airfield, rolling gently from side to side as the pilot puts it through its paces. On the ground, observers watch intently. The jet is brand new, an experimental aircraft being flown for the first time. So far, it has performed magnificently. The designers of this airplane, who are among the observers on the ground, are thrilled.

At the controls of the jet is a seasoned test pilot. He has pushed the new jet to the edge of its capabilities and is now prepared to bring it in for a landing. He takes the jet through a wide turn, then heads back toward the airfield. As the observers watch, he brings it down for what looks to be a picture perfect landing.

But just before he reaches the field, something goes wrong. The jet's wings begin to shudder. All at once it starts to flip upside down. Before it's too late, the pilot ejects—but the jet continues rolling over as it overshoots the airfield and crashes into the desert in a burst of scarlet flame.

Everything seemed to be going perfectly until the pilot started to land. The computer simulations of the jet's flight had shown it performing flawlessly. What went wrong?

It is the year 2419. Astronomers on the planet Earth and elsewhere in the solar system have for several months been observing a peculiar phenomenon. The orbit of the planet Pluto, outermost of the nine planets, has become increasingly unstable. The planet has begun to wobble as it travels around the sun, following a bizarre path through the sky like no other planet the astronomers have ever observed.

The inhabitants of Pluto Base, a complex of scientific laboratories located deep in the ice covering the outermost planet, have been recalled to Earth. Just in time, too. Even as the astronomers watch, Pluto veers away from its normal orbit and shoots off into the deepest reaches of outer space.

What has caused the planet Pluto to fling itself out of the solar system? Has it been captured by some alien force? Or is it merely following the laws of nature?

The chief of the Centers for Disease Control sits behind his desk and ponders a document that has just been placed in his hands. It holds disturbing news. For months, he has been tracking the progress of an epidemic that has struck cities up and down the East Coast of the United States. An emergency vaccination program has been instituted in several major cities, to make a large portion of the population immune to the disease. According to the best computer projections, the vaccination program should have been sufficient to stop the disease in its tracks.

Yet the message he has just received tells him that the disease has broken out again simultaneously in several cities. The vaccination program has been useless. The epidemic is worse than ever.

He shakes his head. Why didn't the computer projections tell him *this*?

A pair of Interior Department wildlife experts stand on a hillside in the Pacific Northwest, looking down on a wide valley. They are making a census, trying to count the population of an obscure species of muskrat. They are worried because, on the basis of figures obtained in previous years, the numbers for this muskrat have been dwindling. The species may eventually be headed toward extinction.

They are actually dreading the numbers that they might obtain in this year's census. If the population of the muskrat has fallen far enough since the previous year, there may be nothing that can be done to save them. This may be the beginning of the end for this species.

But as they begin counting the muskrats in the valley, they make a surprising discovery. The population has increased! There are more than ever before. Furthermore, the numbers have gone up more than they ever dreamed they might. It's as though the population was never in danger at all.

Nothing in the figures from previous years could have prepared them for a result like this!

The doctor stares down benevolently at her patient. The man had entered the hospital several days earlier complaining of pains in his chest. A quick examination revealed that he was having a heart attack.

He was immediately given medicine and attached to monitoring devices that would keep track of his condition. Now, according to the monitoring devices, his condition has stabilized. The patient is ready to be moved out of the intensive care unit. The doctor signals to a pair of orderlies to disconnect the monitoring units and move the patient's bed to his new room. But before the disconnecting begins, an urgent beeping sound fills the air.

Lights flash on the monitors. A jagged line appears on a small oscilloscope screen. The patient is having another heart attack! The doctor gives orders for an emergency team to be sent immediately to the intensive care unit.

As she begins to treat her patient, she wonders why he's taken this sudden turn for the worse. His condition had been improving so!

A husband and wife pack their children into a car on a Saturday morning and drive into the country. It's a beautiful day. The sun is shining, the sky is blue, and the air is warm. Best of all, the weather report on the morning news says that the weather will stay that way all day, with barely a chance of rain.

When they arrive in the country, they find an open field and spread their picnic blanket in the middle of a grassy meadow. The mother opens a container of potato salad and the father breaks out an assortment of cold sandwiches. With delighted smiles, they begin to eat.

Suddenly the sky begins to grow dark. The family looks up worriedly, but they remember the words of the weatherman and continue to eat. Only moments later, rain begins to fall. Before they can pack up the picnic and make it back to their car, they are all soaking wet.

That's the last time we listen to *that* weatherman! they mutter, as they start the long drive back to the city. He should stick to telling bad jokes about his toupee and sending happy birthday greetings to 100-year-old ladies in Iowa. He doesn't know beans about the weather!

Unexpected things happen.

That's a fact of life. Most of us learn to live with the unexpected by the time we are old enough to talk. We reach for a glass of water and spill it on the floor instead of drinking it. We go to the store only to discover that it's closed because of a fire the day before. We watch a football game and see the heavily favored team go down in flames.

All of these vignettes are about the unexpected. They concern events where things did not go as planned, where the real world defied the efforts of experts to predict how it would behave.

Experts don't like the unexpected. A person who has learned everything there is to know in a certain field likes to believe that nothing can surprise him or her. Once you know everything, the experts like to believe, nothing can catch you off guard.

And yet experts are constantly being confounded by surprises. Things happen that they never predicted. Carefully planned projects and experiments go awry. Weather predictions turn out to be wrong. Experimental planes don't behave the way they should. Epidemics take off in surprising directions. Astronomical objects behave in ways that astronomers did not expect them to.

Where do these surprises come from? Why can't experts learn to expect the unexpected?

Traditionally, experts have blamed these surprises on outside influences or imperfect data. Things were more complicated than we realized, the experts say, when something goes wrong. By taking these extra complications into account, we can still learn not to be surprised.

But now scientists, studying the world around us with the aid of powerful computers, are beginning to realize that surprise is inevitable. Systems such as the weather and the stock market have surprise built into them. They will always behave in unexpected ways, no matter how well we understand them. It is in their nature to do things we can't predict.

Scientists even have a name for this kind of unexpected behavior. They call it *chaos*.

In ordinary English, the word *chaos* refers to any kind of disorder or confusion. A messy room or disorderly group of people is said to be "chaotic."

But scientists have a more specific meaning for the word *chaos*. They use it to refer to simple things that behave in complicated and often unexpected ways. And as scientists take a close look at the world around us, they are discovering more and more things that are "chaotic"—that is, that behave in ways that continually surprise us and confound our ability to predict the way that they will behave in the future.

Some of the things that scientists currently believe may be chaotic include the weather, the stock market, populations of living creatures, the orbits of several planets in our solar system, even our very hearts and minds. And, because these things are chaotic, it may not be possible to predict what they will do over long periods. We may never have the ability to make perfect weather forecasts or to know what the stock market will do tomorrow. We may not be able to predict the state of the solar system a million years from now. And certainly nobody will ever be able to predict the capricious workings of the human mind.

But at the same time as scientists are realizing that much of the world is chaotic, they are discovering a strange order in chaos. Things that seem to be disordered and unpredictable actually follow strange and intricate patterns. And by understanding these patterns, scientists hope to understand these so-called chaotic systems in ways that they have not been able to understand them until now.

Chaos theory, as this new area of scientific study is called, is like an old-fashioned good-news bad-news joke. The bad news is that much of the world around us behaves in an unpredictable fashion. The good news is that unpredictable things can be predicted.

Does that sound like a paradox? In a way, it is. Chaos theory is a brand new area of science and scientists are still endeavoring to understand it. According to their findings, na-

A weather map. Meteorologists must compile enormous quantities of data in order to make their daily forecasts. But practitioners of the new science of chaos have wondered whether it will ever be possible to predict the weather accurately.

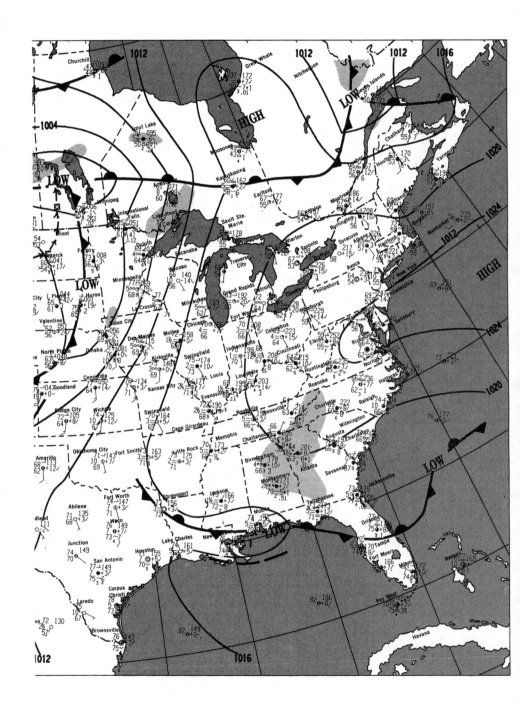

ture is full of unpredictable things that follow predictable patterns—or, at least, *almost* predictable patterns. And these patterns unite many aspects of nature and the human world that were previously believed to be separate, from weather to the human heartbeat to the way airplanes fly and the brain thinks.

Because chaos theory is a new science, it has so far raised more questions than it has answered. Thus, in the pages that follow, we may not be able to answer all of the questions that you may have about this new way of looking at the world around us. But the questions themselves are often as fascinating as any answers could be.

And, because it is the digital computer that has made the chaos revolution possible, we have included several short computer programs in this book. If you have access to an IBM-compatible microcomputer, you can type in these programs and explore the strange world of chaos yourself. Or, if you don't have a computer or simply don't want to take the time to type them, we'll show you what would happen if you did. In the final chapter, we'll take you on a wild microcomputer journey into a new world called the *Mandelbrot set,* where you can see dazzling images of a chaotic universe inside the computer—in full color, right on the computer screen.

But first, let's talk about what science was like before the chaos revolution began, so that we can understand just what it is that's so revolutionary about chaos. Specifically, let's talk about the way that scientists use numbers, and the way that those numbers give scientists a better understanding of the universe.

TWO

Simple Numbers, Simple World

For the last three hundred years, scientists have believed two things above all else.

The first is that it is possible to describe the world in terms of numbers and equations. Everything about the world—and by "world," we mean the entire universe that surrounds us—has a quantity associated with it. And these quantities can be expressed as numbers.

These numbers might represent weight or volume or velocity or some more esoteric property like electric charge or strangeness or charm. (Strangeness and charm in the scientific sense are properties of certain infinitesimally small particles and have nothing to do with the common meaning of the words.) We can say, for instance, that a steak weighs 1.4 pounds or that a car is traveling at 50 miles per hour (mph) or that a subatomic particle has an electric charge of -1. If we know all the numbers that describe a thing, then we know everything that there is to know about that thing.

Equations, in turn, allow scientists to show relationships between the various numbers that they use to describe the world. When scientists plug the appropriate numbers into an equation, they can perform mathematical calculations that allow them to deduce new numbers, numbers that they didn't previously know about. And these new numbers tell them things about the world.

For instance, here's a familiar equation from elementary physics:

$$\text{Distance} = \text{velocity (rate)} \times \text{time}$$

What this means is that the distance a moving object travels is equal to the time spent traveling multiplied by the object's speed. That's not as complicated as it sounds. In fact, you've probably used this equation in your head without even realizing it. For instance, suppose that you're traveling in a car at a speed of 50 miles per hour. If you continue moving at this speed for 3 hours, how far will you travel? Right! You'll have traveled 150 miles. Although you may not have thought that you were solving an equation when you figured that out, you were actually plugging numbers into the equation to replace the words *velocity* and *time*. The resulting equation, written down, looks like this:

$$\text{Distance} = 50 \times 3$$

where 50 is the velocity of the car and 3 is the number of hours traveled. Since 50 times 3 is 150, the car will travel 150 miles in 3 hours.

In an equation like this one, the words *distance, velocity,* and *time* represent *variables*: that is, numbers that can take on different values depending on the specific situation to which you are applying the equation. (They're called variables because they can *vary*.) You can use just about any possible value for the variables velocity and time and still get a valid answer when you solve the equation. The value of distance, on the other hand, is determined by the values of velocity and time. You don't need to know the value of distance in order to solve this equation. In fact, if you already knew the value of distance, you wouldn't need to solve this equation at all. The whole point of the equation is to find the distance that an object, such as a car, travels when you already know the time and the velocity.

If you already know the distance that you're traveling and the velocity you'll be traveling at but need to know the time it will take, you can turn this equation around into a slightly different form:

$$\text{Time} = \text{distance/velocity}$$

This is the same equation, though it's been manipulated slightly by the rules of the form of mathematics. (If you haven't studied much math yet, don't worry about what those rules are. Just take our word for it that this equation is the same as the equation distance = velocity × time.) Now you can insert values for distance and velocity and learn the value of time. Suppose, for instance, that you are traveling from your hometown to another town 220 miles away at a constant speed of 55 mph. How long will the trip take? When we put these values in place of the variables distance and velocity, we get:

$$\text{Time} = 220/55$$

When we divide 220/55 we get 4, so the trip will take 4 hours.

If for some reason you happened to know how long it took to get to some place and how far away the place was, you could rearrange this equation one more time so that it would allow you to calculate your velocity in terms of the distance and time. The equation would now look like this:

$$\text{Velocity} = \text{distance/time}$$

Suppose that you had taken a 100-mile trip and had arrived at your destination after traveling 4 hours. You want to know what your average speed was for the trip. You could plug these values into the equation to get

$$\text{Velocity} = 100/4$$

Dividing 100 by 4 gets us 25, so you now know that you must have traveled at an average speed of 25 mph.

This, of course, is only one of the equations that scientists use to describe the universe around us, and it's one of the simplest ones, at that. A brief perusal of a stack of science textbooks will tell you that scientists have numbers and equations for describing the life and death of stars (not to mention the life and death of the universe itself), the growth of populations of animals, the flow of water, the way in which light travels through a lens, the speed at which objects fall, the interaction of subatomic particles—and just about anything else you can think of that a scientist might conceivably be interested in.

Some scientists even believe that it may be possible to devise one set of equations that will describe the entire universe. Once we know those equations, we will know everything there is to know about the universe. In theory, we could apply the rules of mathematics to those equations and rediscover all of the other equations that scientists have learned over the years, and quite a few that they haven't. In practice, however, this probably won't be feasible.

Most scientists don't worry about describing the entire universe in this way. Instead, they worry about the small portion of the universe that they have chosen to study. Scientists who study stars, for instance, worry only about those equations that describe stars. Scientists who study populations of animals worry only about those equations that describe populations of animals. Scientists who study the interaction of subatomic particles worry only about those equations that describe the interactions of subatomic particles. And so on.

The important thing about describing the universe in terms of equations is that, once a scientist knows the numbers and equations that describe the part of the universe that he or she is studying, he or she can then solve those equations on a piece of paper without even leaving the comfort of the office. It's like capturing a piece of the universe on paper where it can be studied easily, without the need for expensive telescopes or particle accelerators or whatever other instruments a scientist might need to study the part of the universe in which he or she is interested.

As we'll see in a moment, an even more important advantage of describing the universe in equations and numbers is that those equations and numbers can be *programmed* into a computer and the computer can solve the equations for the scientist. The advent of the digital computer roughly fifty years ago made it possible for scientists to take a small portion of the universe and create a mathematical model of it inside the computer. The scientist can then study this model in ways that he or she could *never* study the real world. A scientist studying the ways that stars evolve can make a mathematical model of a star explode inside the computer. A scientist studying subatomic particles that would normally be too small to see can watch mathematical models of those particles inside the computer. And so forth.

Of course, the scientist must remember that the exploding stars and interacting particles inside the computer aren't real. They are only *simulations*, based on equations that the scientist has programmed into the computer. It is necessary for the scientist to compare the results of these simulations with the real world as often as possible, to make sure that the things that are happening inside the computer can happen in the real world too.

The *other thing* that scientists have believed for the last 300 years is that the equations that describe our universe are basically simple. When all of the equations (or, perhaps, the single equation) that govern our universe are known, scientists believe that they will turn out to be simple equations—simple, at least, to those scientists who understand the sophisticated mathematics on which they are based.

It would seem to follow, then, that the world itself must be a pretty simple place, since it is described by those simple equations. Although the world may look complex to those of us who are not scientists, it actually must have an underlying simplicity that is obvious to the scientists who understand the equations that make it tick.

There is plenty of reason to believe that this is true. Whenever scientists have learned new equations to describe the universe, they have discovered that the universe—or at least certain aspects of it—is simpler than might at first have been imagined.

Consider the equation we introduced earlier in this chapter. It can be used to calculate the time, distance, or velocity of any car trip that you might take. In this sense, at least, all car trips are alike. You don't need a different equation for different car trips. Think of how confusing that would be! This one equation will do for all.

Furthermore, that one equation can be used to calculate the time, distance, and velocity of an airplane, a train, or a spaceship traveling between Earth and the moon, as well as a car traveling on the surface of the Earth. It ties all of these things together, showing that there is an underlying simplicity to the way that they travel.

This may seem obvious to you. But, in the seventeenth century, the English physicist Sir Isaac Newton discovered something that wasn't nearly as obvious. Newton worked out the equations that described the way in which an apple (or any other object) fell to the surface of the Earth and the equations that described the way that the moon orbited around the Earth—and discovered that they were the same equations. Although it may be obvious that an equation such as distance = velocity $\times$ time describes a spaceship as well as an automobile, it isn't nearly as obvious that the equations describing a falling apple also describe the moon. When Newton discovered that this was so, it told him something important about the moon: that the moon was like a giant apple, falling eternally around the Earth.

A computer simulation of
a head-on collision of
two neutron stars.
Scientists generated this image
by using a supercomputer to
solve the equations of
Einstein's general theory
of relativity.

And when Newton proceeded to work out the equations that described the way that the planets, including Earth, orbited the sun, he discovered that they were the same equations that described the moon and the apple. The entire solar system could be described in a few simple equations. What had seemed to generations of scientists to be complicated was, in fact, surprisingly simple.

Since the time of Newton, scientists have made many similarly startling discoveries. Scientists who studied the nature of matter, for instance, learned that the tens of thousands of different substances known to science could be explained by the numbers and equations describing only about a hundred different objects called atoms. And scientists studying these atoms learned that the hundred or so different atoms could be explained by the numbers and equations describing three different objects called protons, neutrons, and electrons.

Each time scientists boiled down some aspect of the universe to a few simple equations, the universe seemed to grow simpler. There was every reason to suspect that the universe would turn out to be a simple place indeed.

But the universe has refused to become as simple as scientists would like it to be. Often, when scientists are studying seemingly simple aspects of the world around us, unexpected complications crop up. Scientists studying the flow of air around the wing of an airplane have to deal with turbulence, where the flowing air stops flowing smoothly and starts flowing in complicated patterns that can render the airplane unflyable. Scientists studying the way that moons orbit around planets and planets orbit around the sun have to deal with strange wobbles in those orbits, wobbles that don't seem to be accounted for in the simple equations that describe those orbits. Scientists studying the way that populations of animals grow and shrink over time find inexplicable patterns in the changing sizes of those populations.

Traditionally, scientists have ascribed these unexpected complications to *noise*, outside interference from other phenomena that don't have anything to do with the phenomenon that is being studied. If scientists can just remove all of the noise from their experiments, so that they can study phenomena such as water flow or planetary orbits in complete isolation from all interference, they believe that the complications will go away. The systems being studied will behave in the simple manner implied by the simple equations that describe them.

But in the last few decades, scientists have acquired a new tool for analyzing the simple equations that they use to describe the world around them. This new tool is the computer. With the aid of this tool, they have discovered that the complications in their experiments are not coming from noise. They are coming from the simple equations themselves. These simple equations are producing much more complicated results than anyone expected they would. And it is from these complicated results that the behavior scientists refer to as chaos arises.

How can simple equations produce complicated results? That's a question that really requires a computer to answer. So, in the next chapter, we'll call on a computer for assistance, just as the scientists themselves have. And in the chapter that follows, we'll take a look at how strange and unexpected patterns can flow out of simple numbers and equations, and how they can produce striking and informative pictures on a computer screen.

C   H   A   P   T   E   R

THREE

Numbers in the Fast Lane

Computers work well with numbers. In fact, a computer won't work with anything else.

A computer is a number-manipulating machine. Put numbers into a computer and it will perform just about any operation on those numbers that you can imagine. It will add them, subtract them, multiply them, and divide them. It will even perform operations that you've probably never heard of, such as rotates and shifts, logical ANDs and exclusive-ORs.

This makes the computer the ideal tool for scientists. If the universe can be described in terms of numbers, then those numbers can be fed into a computer. And the computer can manipulate those numbers in a way that will tell us something new about the universe.

Before a scientist can put numbers into a computer, however, he or she must program the computer: that is, tell it what it's expected to do with those numbers. A computer program is simply a list of instructions, written in a language that can be "understood" by a computer, telling the computer what to do.

A computer program for modeling the real world usually

consists of one or more equations that describe the real world, plus some numbers that can be plugged into those equations. (Sometimes these numbers are typed into the computer at the time the program is run. At other times, they are actually put into the program itself.) The *output* of the program, the part that you see when the program is executed by the computer, usually consists of either lists of numbers or a graph, such as you might draw on a piece of graph paper.

This may sound pretty dull, but it can give the scientist a lot of important information. And, later in this book, we'll see some graphs on the screen of the computer that are anything but dull.

Let's look at an example of a computer program that contains an equation and tells us some useful things about the real world. This program is written in a computer language called BASIC. If you have access to a microcomputer, you might want to type this program into the computer yourself and run it to see whether the results you get agree with the results in this book. (See Appendix A at the end of this book for instructions on how to type programs in the BASIC programming language.)

Remember the equation in the last chapter for calculating distance in terms of velocity and time? Here is that equation again, only this time it is written in the computer language called BASIC:

$$D = V * T$$

You'll notice that we've shortened the names of the variables to D, V, and T for distance, velocity, and time, because some versions of the BASIC programming language don't like variable names longer than one or two letters. And, in case you're wondering, the asterisk (*) is the symbol used by most computer programming languages to represent multiplication, because the traditional multiplication sign might be confused with the letter X.

Of course, this simple equation doesn't represent a complete computer program. We also need a way of telling the computer what numbers we want it to plug in to represent the velocity and the time. And we need a way for the computer to print the resulting distance someplace where we can read it. Here's a complete program that will do all of those things. (Be careful when you type programs into your computer. Even a small typographical error could cause the program to act in a very peculiar manner.)

```
10   INPUT "Velocity (in miles per hour)";V
20   INPUT "Time (in hours)";T
30   D=V*T
40   PRINT "The distance traveled is ";D;" miles."
```

If you ran this program on a microcomputer, it would first print the words "Velocity (in miles per hour)?" on the video screen of the computer and wait for you to type a number. (The question mark after the words "Velocity (in miles per hour)?" is added automatically by the computer, even though we didn't specify it in the program.) You would then type a number and press the key marked ENTER or RETURN on the computer's keyboard. It would then type the words "Time (in hours)?" on the video screen and wait for you to type a number. Finally, after you'd typed both numbers, it would print out the number of miles that you've traveled.

Let's give the program a try. To make the computer execute a BASIC program once we've typed it in, we type the word RUN, then press the key marked ENTER. (If you don't have a computer or don't want to type in this program right now, just follow along as we describe what happens.) Now we'll tell the computer that our velocity is 50 mph:

Velocity (in miles per hour)? 50

And we'll tell it that we'll be traveling for 7 hours:

Time (in hours)? 7

The computer responds by typing:

The distance traveled is 350 miles.

Bingo! The computer has performed the calculation for us and given us the result! You can easily verify the computer's math by multiplying 7 times 50 in your head. And you should get the same answer: 350.

In fact, you can verify the computer's math so easily that you might wonder why we bothered having the computer do it in the first place. You could have figured this one out in your head just as quickly as the computer did, without all the hassle of typing in a computer program.

That's true. But we've barely scratched the surface of what the computer can do. Although computers are good at solving quick equations like this one, what they are *best* at is solving the same equation over and over again, using different numbers each time. The computer never gets bored with solving the same equation repeatedly, and it can do it with a speed and accuracy that few, if any, human beings can duplicate.

Let's revise our computer program so that, instead of asking us to type in numbers for velocity and time, it produces its own numbers. Here, for instance, is a program that will calculate how far you would travel if you were moving at 50 mph for 1, 2, 3, 4, 5, 6, 7, 8, 9, or 10 hours:

```
10   FOR T = 1 TO 10
20   D = T*50
30   PRINT"IF YOU TRAVEL FOR ";T;" HOURS YOU WILL
     GO ";D;" MILES."
40   NEXT T
```

If you run this program on your computer, it will print the following on the video screen:

```
IF YOU TRAVEL FOR 1 HOURS YOU WILL GO 50 MILES.
IF YOU TRAVEL FOR 2 HOURS YOU WILL GO 100 MILES.
IF YOU TRAVEL FOR 3 HOURS YOU WILL GO 150 MILES.
IF YOU TRAVEL FOR 4 HOURS YOU WILL GO 200 MILES.
IF YOU TRAVEL FOR 5 HOURS YOU WILL GO 250 MILES.
IF YOU TRAVEL FOR 6 HOURS YOU WILL GO 300 MILES.
IF YOU TRAVEL FOR 7 HOURS YOU WILL GO 350 MILES.
IF YOU TRAVEL FOR 8 HOURS YOU WILL GO 400 MILES.
IF YOU TRAVEL FOR 9 HOURS YOU WILL GO 450 MILES.
IF YOU TRAVEL FOR 10 HOURS YOU WILL GO 500 MILES.
```

The computer has tirelessly calculated exactly how far you will travel if you go at 50 mph for 1 hour, 2 hours—right on up to

10 hours. Furthermore, if we change the first line of the program to the following:

```
10 FOR T=1 TO 100
```

the computer will calculate your mileage right up to 100 hours. And if you change that 100 in the first line to 1,000, it will calculate your mileage up through 1,000 hours. And so forth. The computer does this quickly and without complaining. And it does it much more quickly and accurately than a human being could.

Once more, however, you might wonder why anybody would bother putting such a simple equation on the computer. What do we learn from it?

Well, we actually do learn something from running this program. But it's difficult to see what we've learned just by looking at the mass of figures that the program puts on the screen. It would be easier if the program drew us some kind of picture.

Computer pictures are usually referred to as *graphics*. And the particular kind of picture that we want the computer to draw right now is called a *graph*. (Don't get these two words confused. Although they are similar and they both refer to pictures that can be drawn on a computer, they don't mean the same thing. A graph, for instance, can also be drawn on a sheet of paper.)

What is a graph? To a scientist, a graph is a drawing that shows relationships between numbers. Because equations also show relationships between numbers, any equation can be turned into a graph. Why would we turn an equation into a graph? Because the important relationships between the numbers are often easier to see in a picture than in a list of numbers written on a page or printed on the screen of a computer.

When a scientist (or anyone else) wants to make a graph, he or she usually begins by drawing two lines on a piece of paper. One of these lines is horizontal (that is, it runs from left to right) and one of these lines is vertical (that is, it runs up and down). The horizontal line is called the *x axis* and the vertical line is called the *y axis*. The two lines should cross each other at some point in the drawing. (See Figure 3.1.)

You can imagine each of these lines as a kind of ruler or yardstick. Like a ruler, each line has numbers marked off on it, beginning with 0 at the point where the two lines cross each other. (Sometimes the 0 isn't marked, since there may not be enough room where the two lines cross to mark it down.) Additional numbers are marked at even intervals moving away from the crossing point, starting with 1 and then 2 and then 3, and so forth. Numbers on the horizontal line to the right of the point where the lines cross are positive numbers and numbers to the left of the point where the lines cross are negative numbers (that is, numbers with minus signs in front of them).

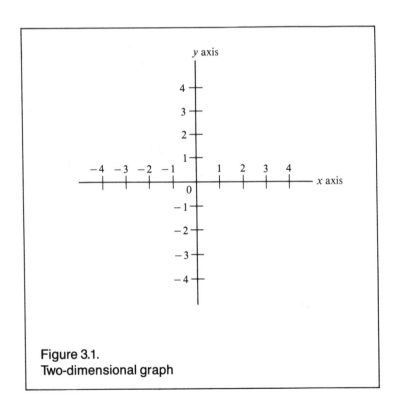

Figure 3.1.
Two-dimensional graph

Numbers on the vertical line above the point where the lines cross are positive numbers and numbers below this point are negative numbers. Only those numbers that are actually needed in the equation being graphed are included. The rest are assumed to be somewhere off the edge of the drawing. Frequently, the negative numbers aren't included at all.

How do you graph an equation? Well, on a *two-dimensional graph* like the one we just described, the numbers on the x axis and the y axis represent the values of two variables in the equation. One of these is a variable that we can change, and the other is a variable that changes when we change the value of the other variable. For instance, we could graph the equation $D = VT$. (You'll notice that we've left out the multiplication sign altogether this time. When you see two variables placed next to each other in an equation like this, it means that they are to be multiplied times one another.) The x axis could represent the variable T and the y axis could represent the variable D. We could then plug in various values for T and the value of D would change. And we could represent these changing values as dots and lines on the graph.

To graph this equation, we would start out by substituting possible values for T and then calculating the resulting value of D. We would then find the point on the graph that represents those two numbers by finding the number for T on the x axis, finding the number for D on the y axis, and then the point on the graph where those two numbers come together. (See Figure 3.2.) We then put a dot at that point.

After repeating this process for several different values of T, we have several dots on the graph. We then complete the process by connecting the dots with lines, just like an old-fashioned connect-the-dots puzzle. (See Figure 3.3.)

But graphing can get tedious pretty quickly. Why go to all this effort when we have a computer to do the job for us?

The following program will create a graph for the equation $D = VT$, using the same numbers that the last program printed out on the screen. If you're going to type this one yourself, you'll need an IBM-compatible microcomputer. It may not

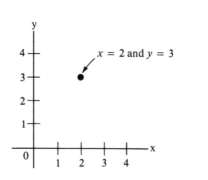

Figure 3.2.
A point plotted

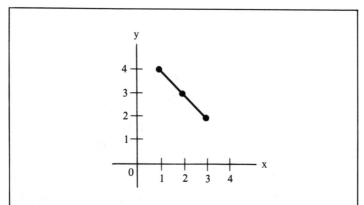

Figure 3.3.
A graph of several plotted points

work with the versions of BASIC supplied with other computers. Here's the program:

```
10   DELAY = 1000
20   CLS: SCREEN 9: KEY OFF
30   LINE (0, 0) – (0, 199)
40   LINE (0, 199) – (319, 199)
50   FOR I = 1 TO 10
60   LINE (0, 200 – (I * (50/3))) – (2, 200 – (I * (50/3)))
70   LINE (I * 25, 199) – (I * 25, 197)
80   NEXT I
90   FOR T = 1 TO 10
100  D = 50 * T
110  PSET (T * 25, 200 – D/3)
120  FOR I = 1 TO DELAY
130  NEXT I
140  NEXT T
150  FOR T = 1 TO 10
160  D1 = (T – 1) * 50
170  D2 = T * 50
180  LINE ((T – 1) * 25, 200 – D1/3) – (T * 25, 200 –
     D2/3)
190  FOR I = 1 TO DELAY
200  NEXT I
210  NEXT T
220  A$ = INPUT$(1)
```

Whew! That's a bit more complicated than the programs we've
looked at up until now! (Once again, be careful when you type
it into your computer. Even a small typographical error could
cause the program to act in a very peculiar manner.)

Most of the new complications, however, are concerned
with getting the picture on the screen. If you look at the line
labeled 100 in the program, you'll still see our familiar equation $D = VT$ (except that we've plugged a value of 50 directly
into the program in place of V, because we don't want it to
change while the program runs). Don't worry about how the
rest of the program works, unless you already have an interest
in BASIC programming.

When you run this program, you'll see the x and y axes
(plural of *axis*) drawn along the left and bottom sides of the
screen. Small tick marks along each axis represent the num-

bers from 1 to 10, starting where the axes meet in the lower left-hand corner. (The actual numbers from 1 to 10 don't appear on the screen because of difficulties putting numbers on a computer screen alongside graphic images.)

After drawing the axes, the program puts a series of dots on the screen, then connects them with lines. These dots represent the values of T and D in the equation $D = VT$. So that you can see the graph being drawn, the program puts the dots and lines on the screen in slow motion. (Different computers run at different speeds. If the graph is drawn either too slowly or too quickly for your taste, you can change the value of the variable DELAY in line 10 to make it go faster or slower. Changing the number after the equals sign [=] in this line to a lower number will speed up the program. Changing it to a higher number will slow down the program.)

If you don't have a computer to run this program on, you

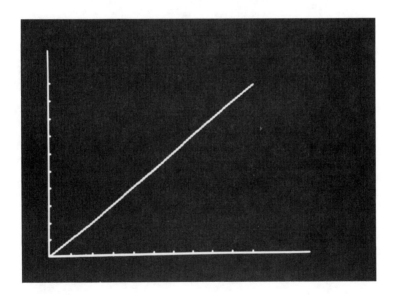

Figure 3.4.
Graph of relationship between
distance, time, and velocity

35

can see a picture of the graph in Figure 3.4. And what do we learn from this graph? We learn that, when the value of T increases in this equation, the value of D increases too. And they both increase steadily, as steady as the straight line on the screen of the computer.

This means that, the longer you drive, the farther you go. This probably doesn't come as a surprise. It would be surprising if you went a *shorter* distance the longer you drove. And yet there are equations that would have produced exactly this result. Some equations, for instance, produce a curved line that rises, then falls. When we see such a line on a graph, we know one variable doesn't always increase as the other variable increases.

What good does it do us to know this? In the next chapter, we'll look at another equation and another graph. This graph will do some very unexpected things. In fact, this graph will give us a look at chaos itself.

FOUR

Simple Numbers, Complicated World

In 1971, an American biologist named Robert May was studying the way that animal populations grow and shrink.

This is a question that has interested scientists since the eighteenth century, when it was first tackled by an English economist named Thomas Malthus. Malthus realized that animal populations, including populations of human beings, could not keep growing indefinitely. Eventually, the population would grow too large for the food supply to support and would begin to shrink as the individual animals starved to death.

Thus, if we were to graph the way that an animal population grows and shrinks, it would form a curve. (See Figure 4.1.) It would go upward rapidly as the population grew larger and larger, then would slow as the food supply became too small to support it. Finally, it would go downward as the population began to starve. Malthus feared that this was to be the fate of human populations, a prophecy that has occasionally proved true in economically disadvantaged parts of the world. Fortunately, human populations have the intelligence not only to slow their birthrates before disaster intervenes but

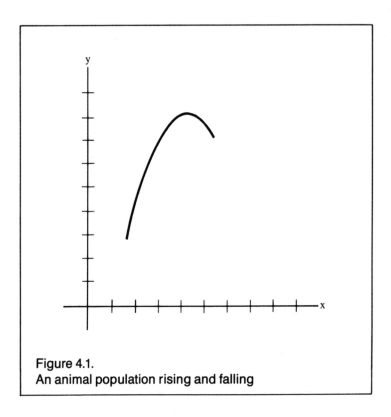

Figure 4.1.
An animal population rising and falling

to develop better and better food production technology to stave off famine. Populations of other animals, however, frequently grow too large for their environment to support and begin to die off.

The curve rarely goes down forever. If it did, the animals would become extinct. More often, as the animal population shrinks, it eventually reaches a size that can be supported by the environment. The food supply is no longer inadequate. The population begins to grow—and the process starts all over again. The curve stops going down and goes back up again instead. (See Figure 4.2.)

These were the ideas that Robert May was studying in 1971. But he was not studying actual animals or actual pop-

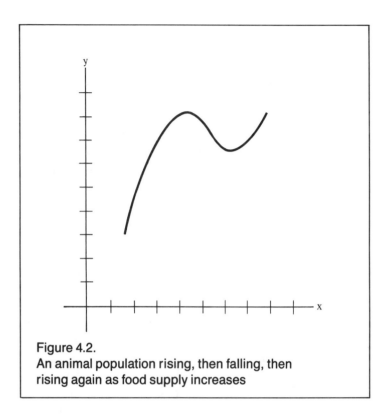

Figure 4.2.
An animal population rising, then falling, then
rising again as food supply increases

ulations of animals. He was studying an equation on a sheet of
paper.

The equation that he was studying is well known in the
sciences. It even has a name. It's called the *logistic difference
equation*. It looks like this:

$$N_{T+1} = RN_T(1 - N_T)$$

That looks pretty intimidating! What do those *subscripts*, the
tiny letters and numbers half a line below the other letters and
numbers, mean? For that matter, what does this whole equation mean?

Actually, it's not as difficult as it looks. The logistic difference equation is an example of an *iterated equation*—an

equation that must be solved over and over again repeatedly to produce meaningful results—and the subscripts T and $T + 1$ simply mean "this time" and "next time" when translated into English. In other words, the value of $N_{T + 1}$ this time you solve the equation will be the value of N_T the next time you solve the equation.

What this equation allows scientists like Robert May to do is to calculate what the size of an animal population will be "next time" on the basis of what it is "this time." The variable N_T represents the number of animals in the population "this time," and the variable $N_{T + 1}$ represents the number of animals in the population "next time." (Actually, the values of these two variables are fractions between 0 and 1 that represent the fraction of the total *possible* population that is currently alive.) The variable R represents the rate at which the animal population increases in size, based on its birth and death rates.

If a biologist is tracking an animal population year by year, he or she can plug in the current population for N_T and the rate of population increase for R, and then can calculate next year's population ($N_{T + 1}$) simply by solving the equation. Then, if the biologist wants to know what the population will be the year after that, he or she can plug in the value of $N_{T + 1}$ for N_T and calculate the size of the population. And the biologist can keep iterating—repeating—this process, watching how the population changes year by year over decades and even centuries.

You might wonder just how accurate this equation is. That's a good question and not an easy one to answer. Population biologists, scientists who study the rise and fall of animal populations, believe that the logistic difference equation accurately models the way that animal populations grow. But it's difficult to test this hypothesis to make sure that it's true. In the real world, animal populations aren't necessarily as simple as this equation makes them seem. There are other factors to take into account than just food supply. There is, in other words, a lot of noise.

For instance, animals don't just die of starvation. They can also die because of severe weather conditions or because they are being preyed on by another species of animal or because of unexpected changes in their environment. And the food supply doesn't necessarily stay the same over time. It can grow and shrink just as the population of animals grows and shrinks. If the animals in the population being studied are carnivores— that is, if they eat other animals—then the population of animals that they eat will grow and shrink as it outgrows its own food supply.

Finally, it's difficult to get accurate population figures for animals. You can't mail out census forms to them the way that the Census Bureau does to human populations. It's necessary for population biologists to go out into the field and actually count the animals one by one—and that's easier said than done.

Despite all of this, population biologists believe that the logistic difference equation represents a pretty accurate model of the way that animal populations would grow and shrink if all the noise, the extraneous influences and inaccurate data, could be removed.

You would figure, then, that the logistic difference equation must show animal populations' changing size in a simple, fairly predictable way. After all, it's a simple equation. And, in Chapter 2, we proposed the notion that a world that can be modeled with simple equations must be a pretty simple world. Any complications must be noise—and there shouldn't be any noise in the logistic difference equation.

That's why Robert May was surprised to discover that the logistic difference equation was very noisy indeed. In fact, when he sat down and began playing with the logistic difference equation on a sheet of paper, he began to get results that were so noisy they made almost no sense at all. Although he didn't have a word for it at the time, what Robert May discovered was chaos.

Let's try solving the logistic difference equation ourselves and see just what it was that Robert May discovered. To make our work easier, we'll use a tool that Robert May didn't have handy in 1971, a computer. (Computers existed in 1971, of course, but the small microcomputers that almost every scientist has on his or her desk today weren't invented until a few years later. Computers in 1971 were big and expensive and were only used for "important" work, not for playing around with simple equations.)

Here's a short program in BASIC that will solve the logistic difference equation for us:

```
10  N=.1
20  INPUT"VALUE OF R";R
30  FOR I=1 TO 30
40  N1=R*N*(1−N)
50  PRINT N1
60  N=N1
70  NEXT I
```

Take a look at line 40 in this program and you'll see the logistic difference equation in a slightly different form. We're now using the BASIC variable N to represent the variable N_T in the original equation and the variable N1 to represent the variable N_{T+1}. (And, of course, we're using the asterisk to represent multiplication.) Otherwise, this equation is the same as before.

Before solving the equation, this program will set the value of the variable N to 0.1. This means that the imaginary animal population that we are going to examine starts out with a population that is only one-tenth as large as the maximum possible population. It then will ask us for a value to set R equal to. Finally, it will iterate—repeat—the equation 20 times, so that we can see how the population changes over the next 20 time periods, printing the population at each time period on the screen. (For the sake of argument, we'll say that these time periods are 1 year long, though if this were a real animal population we would need to use a period based on the way in which the animals actually reproduced their numbers.) After every iteration, it sets the value of the variable N to the value of variable N1 from the previous iteration.

Let's run the program and see how some imaginary populations of animals grow and shrink over time. When it asks for a value of R, give it a value of 2. You'll see a list of numbers run down the screen of the computer, like this:

.18
.2952
.4161139
.4859263
.4996039
.4999997
.5
.5
.5
.5
.5
.5
.5
.5
.5
.5
.5
.5
.5
.5

An interesting thing has happened here. At first, our imaginary population of animals grows, but then the growth slows down. It doesn't start falling, though. It comes to a complete standstill at .5, meaning that the population has grown to one-half of its maximum size, then simply stopped. This happens by the seventh iteration of the equation. After that, every time the equation is iterated again the result is the same: .5. A population biologist would say that our imaginary population has stabilized. It has reached the perfect number and then stopped growing.

This is a rather neat result. The population has turned out to be remarkably predictable. As long as the value of R for our imaginary population is 2, the population will eventually stabilize at half of its maximum size, provided that there isn't any outside noise, such as interference from other species or a drought that destroys the food supply. And not only does it stabilize but it stabilizes remarkably quickly. Maybe this population biology stuff isn't so difficult after all! In fact, maybe predicting the way that animal populations grow is simpler than we ever would have guessed.

When an iterated equation is drawn toward a single number like this and then stays there once it has reached it, mathematicians refer to that number as an *attractor*. For the logistic difference equation, when the value of R is 2, the attractor is .5, as we have seen.

Will this attractor stay the same for other values of R? It's easy enough to find out. Just run the program again, and try out different values of R. You are warned, however, not to try values of R that are greater than or equal to 4, because they'll create bugs in the program. (This is a technical problem, which we're not going to explain further here. Just take our word for it that it's true.)

Let's run the program and give it a value of 1.5 for R. Something similar, though not exactly the same, happens. This time, the population grows for 26 periods before it settles down on an attractor. And this time the attractor is .333333 . . . instead of .5. (The ". . ." after the "333333 . . ." means that the number would keep repeating forever if we kept writing down digits. If you wrote this number as a fraction, it would be ⅓, but it's impossible to write it out fully as a decimal.) This means that the population would grow to ⅓ its normal size, then stabilize.

If we run the program using 2.5 for R, the same thing happens again. This time, the population shoots up to a high of .61474 in just three periods—and starts to shrink. But it immediately starts to grow again right afterward and proceeds to shrink and grow for more than 20 periods. This is pretty much the way we assumed that such a population would act when we talked about populations of animals at the beginning of this chapter. But then it settles down to an attractor of .6: that is, the population stabilizes at six-tenths of its maximum possible size.

Before we try some other numbers in this program, let's make a guess as to what it will do. So far, all numbers larger than 1 have produced populations that grow (and sometimes shrink) before stabilizing—that is, settling down at an attractor. (Numbers smaller than 1 tend to produce populations that become extinct: that is, they rapidly shrink until they reach 0. In a sense, you could say that 0 is the attractor for these values of R.) Furthermore, the larger the value of R, the larger the attractor at which the population settles down. So as we give larger and larger values to R, it seems reasonable to expect the population to stabilize at higher and higher attractors.

You'd think this would make the job of population biologists pretty easy. All they have to do is find the value of R for a population of animals and they'll know at exactly what percentage of the maximum population the animals will eventually stabilize. A graph of the attractors for this equation should show a steadily rising line or curve that goes higher as the value of R goes higher. What could be simpler? Of course, in practice a population biologist has to deal with other variables like the ones we discussed earlier, such as predators and problems with the food supply, but that's just noise. The underlying equation that governs population size is just as simple as could be.

Or is it?

So far, we haven't tried the logistic difference equation with values higher than 2.5. (And, as we noted earlier, we will not be trying it with values greater than or equal to 4.) So what happens when we use values in the range 3 through 3.99?

Strange things indeed. It's in this range that the simplicity of the logistic difference equation falls apart—and we enter the realm of chaos.

Run the program again and try a value of 3.1 for R. Now something slightly different happens. The population grows for a few periods, then it starts shrinking and growing for quite a few periods before it settles down. And when it finally settles down, it doesn't settle down to a single attractor. It settles down to *two* attractors, .5580143 and .7645665, switching back and forth between them. This means that the population will be roughly .55 of its maximum possible size for one period, then .76 of its maximum possible size for the next period, and will alternate back and forth between these two sizes forever.

When a system that has a single attractor goes to two attractors when the value of a variable is increased, mathematicians say that the system has *bifurcated,* a word that means roughly "split in two." And when a system starts bifurcating, it often doesn't know where to stop.

Try using higher and higher values for R (without going over 3.99) and see what happens. When you get to values larger than 3.4, the system bifurcates again. Now there are four attractors. And as the value of R goes higher, it will bifurcate again and again, with 8 attractors and 16 attractors and even 32 attractors. Will it ever stop bifurcating?

Yes. When you get to values higher than 3.57, there are no attractors at all. The population never settles down. It just jumps from one number to another every period, seemingly without rhyme or reason. The population has become completely unpredictable.

In fact, the population has become chaotic.

A population biologist faced with such a population would have almost no way of predicting what the size of the population would be in the next period because it would simply jump about in a seemingly random way. Yet the changes in size of the population aren't really random. They are changing according to the simple logistic difference equation, which can be solved by computer. And yet even small changes in the value of R now cause big changes in the way the population grows and shrinks. A biologist faced with such a population might throw up his or her hands in despair.

More precisely, a biologist faced with such a population might be tempted to blame this chaotic behavior on noise: outside influences that are fouling up the simple equations that govern the growth of the population. But there's no noise in our computer program and the chaotic behavior is still there.

The chaotic behavior is in the equation itself. No noise is necessary.

And yet a biologist who looks at this chaos and sees no order in it at all would be just as wrong as a biologist who expects the population to change in simple, predictable ways. There *are* strange types of order in this behavior. To see them, we'll need to draw a picture on the screen of a computer. The following program, which runs in BASIC on IBM-compatible computers, will paint an astonishing picture of chaos in action:

```
10 KEY OFF:CLS
20 SCREEN 7
30 DEFSNG A-Z
40 COLUMNS = 320
50 ROWS = 200
60 START = 1
70 FINISH = 3.999
80 TOP = 0
90 BOTTOM = 1
100 MAXREPS = 10
110 HEIGHT = BOTTOM - TOP
120 VPCT = 1/HEIGHT
130 FOR R = START TO FINISH STEP
    (FINISH - START)/COLUMNS
140 X = .1
150 FOR I = 1 TO 100
160 X = R * (X - X * X)
170 NEXT I
180 FOR I = 1 TO 30
190 X = R * (X - X * X)
200  PSET ((R - START) * COLUMNS/(FINISH - START),
     ROWS-(X - TOP) * ROWS * VPCT)
210 NEXT I
220 NEXT R
230 A$ = INPUT$(1)
```

This program is written to draw graphics on the computer screen using what is called the IBM CGA graphics adapter. It should work on any IBM-compatible computer with *any* kind of graphics adapter except Hercules graphics adapters. However, if you are using the higher-resolution EGA graphics adapter, you'll want to make the following changes to take full advantage of your computer's graphic capabilities:

```
20 SCREEN 9
40 COLUMNS=640
50 ROWS=350
```

And if you have the even higher-resolution VGA graphics adapter, you'll want to make these changes instead:

```
20 SCREEN 12
40 COLUMNS=640
50 ROWS=480
```

If you make these changes and get an "Illegal Function Call in Line 20" error when you try to run the program, it may be that the version of BASIC that you are using doesn't know what graphics adapter you are using. Unfortunately, this means that you'll have to go back to a lower-resolution version of the program. If this happens, or if you aren't sure what graphics adapter you are using, experiment with these different versions until you find one that works. (And, once again, be on the lookout for typographical errors that will prevent the program from working properly.)

What this program does is draw what mathematicians call a *bifurcation diagram* on the screen of a computer. This bifurcation diagram is based on the logistic difference equation, which you'll find in lines 160 and 200 of the program. Each vertical line of dots on the computer's screen represents a possible value of R between 1 and 3.99, with 1 on the far left of the screen and 3.99 on the far right. The dots that are drawn on these lines represent the values that the variable N can take after it's settled down to an attractor or attractors (or failed to settle down to an attractor).

This will be a lot clearer if you actually run the program (or look at the pictures in this book of what would happen if you did). When the program is first run, you'll see a black background with a white line slowly snaking up from the lower left-hand corner of the screen. (See Figure 4.3.) This white line is the attractor for the logistic difference equation, slowly

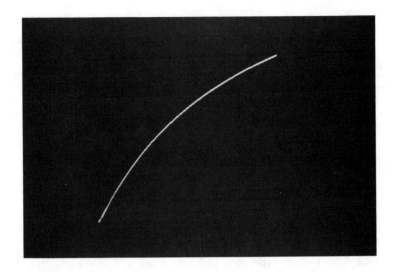

Figure 4.3.
A single attractor

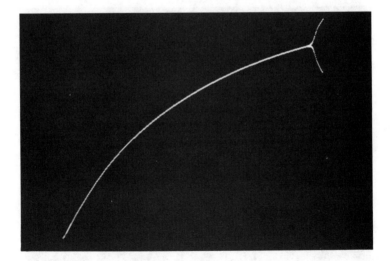

Figure 4.4.
The attractor bifurcates.

getting larger as the value of R increases, just as we noticed it did earlier. The white line rises very steeply at first, then begins to level off, as the value of the attractor rises less steeply.

So far, so good. But when the line is two-thirds of the way across the screen, it starts to bifurcate. (See Figure 4.4.) You can see now why mathematicians use a word that means "splitting in two" to describe what happens to the attractor at this point. The line literally divides in two. One line then keeps going up, almost like a miniature version of the line from which it split. The other line goes down. And then, suddenly, both of these lines split in two, like the glowing streamers in a fireworks display. (See Figure 4.5.) Now there are four lines, following exactly the same patterns as the two lines from which they split. And then the four lines split into eight lines and the eight lines into sixteen and then—chaos! (See Figure 4.6.) Dots all over the screen, like a shimmery veil lowered

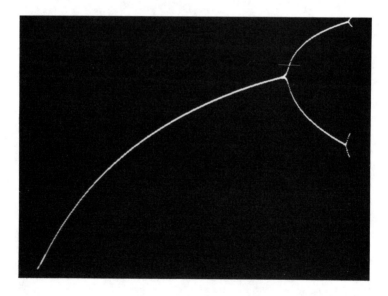

Figure 4.5.
Now there are four.

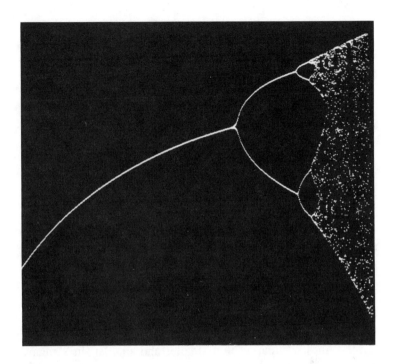

Figure 4.6.
Chaos!

from above. But even these dots seem to be caged inside lines like the ones that form the first part of the diagram. You can even see the faint ghost of lines running through the chaotic dots, as though the diagram almost wants to settle down on an attractor and can't quite manage it.

And what are those strange empty areas in the middle of the chaos? If you look closely (in the high-resolution versions of the program, at least), you can see three lines crossing the empty area: one at the top, one at the bottom, one in the middle. It's as though, right in the middle of chaos, the diagram suddenly settles down to three attractors—then immediately returns to chaos again.

The most amazing thing about the bifurcation diagram is the amount of detail in it. By changing a couple of numbers in this program, we can "zoom in" on various parts of the diagram and take a look at that detail.

For instance, take a look at Figure 4.7. It shows the complete bifurcation diagram with a small box around one of the points at which chaos begins. Let's zoom in and take a closer look at that part of the diagram. Press a key to stop the program and make the following changes to it:

```
60  START=3.39
70  FINISH=3.69
80  TOP=.3166
90  BOTTOM=.4166
```

(From now on, we'll put the values of START, FINISH, TOP, and BOTTOM underneath the illustrations and you can change those values in the program to zoom in on these areas.)

This time, when you run the program you'll see a magnification of the part of the screen that was in the box in Figure 4.7. For a picture of this magnified area, look at Figure 4.8. Why, it looks just like another bifurcation diagram! You can see the lines bifurcating into smaller and smaller lines, many of which were too small to see in the first diagram. Yet the lines follow somewhat different patterns than they did the first time.

Now let's zoom in on the area surrounded by the box in Figure 4.8 and see even more detail. Set START, FINISH, TOP, and BOTTOM to the values shown under Figure 4.8 and run the program again. Darned if it doesn't look like yet another bifurcation diagram! (See Figure 4.9.) The lines are still splitting, into smaller and smaller lines. But now you can also see some small differences from the original bifurcation diagram.

Time to zoom back out. Figure 4.10 shows the full bifurcation diagram with a different area surrounded by a box. This time, the box is in one of those strange empty areas inside the chaotic zone. When we zoom in on this area we find—another bifurcation diagram (See Fig. 4.11)! Right in the middle of the chaotic area there is suddenly another set of lines bifurcating in exactly the same way that the lines in the first part of the

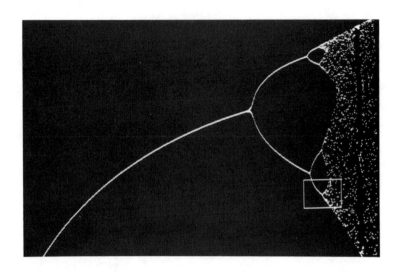

Figure 4.7.
Notice the box.

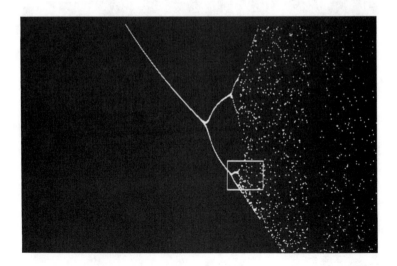

Figure 4.8.
Magnification of the box

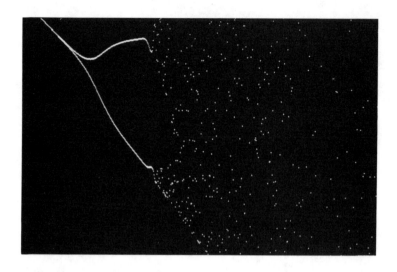

Figure 4.9.
Another level of magnification (START=3.5601,
FINISH=3.59, TOP=.34, BOTTOM=.35)

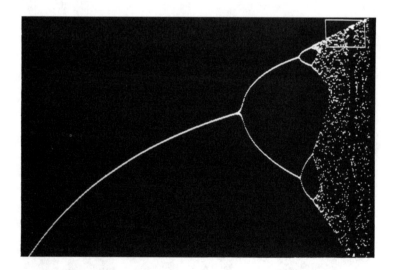

Figure 4.10.
Now the box is in the upper
right-hand corner.

diagram did! Where did this strange order in the middle of chaos come from?

No one is quite sure. Mathematicians have shown that as the value of R goes up in the logistic difference equation, there will be ordered areas like this in the middle of chaotic areas. But it's difficult to say why they are there.

After Robert May wrote a paper on the chaos that he discovered in the logistic difference equation, a team of mathematicians wrote an abstruse mathematical paper on the bifurcation diagram called "Period Three Implies Chaos." This was the first use of the word *chaos* to describe what happens with the logistic difference equation. It was coined by the American mathematician Jim Yorke.

The logistic difference equation isn't the only one that produces this diagram. In fact, there are many other equations

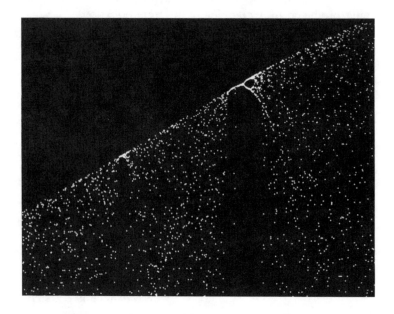

Figure 4.11.
Bifurcation in the midst of chaos. (START=3.6536, FINISH=3.9526, TOP=.8833, BOTTOM=.9833)

that will produce bifurcation diagrams that look just like this one. One of them is:

$$Y = R \times SIN (X)$$

(SIN is a special operation known as a transcendental function. Don't worry about what it does.)

To see the bifurcation diagram based on this equation, you need only make the following changes to the program:

```
 60 START=1
 70 FINISH=3.99
 80 TOP=0
 90 BOTTOM=4
160 X=R*SIN(X)
190 X=R*SIN(X)
```

When you run this version of the program, you'll see a diagram like the one in Figure 4.6. Wait! Isn't that the same bifurcation diagram as before? Yes, it is; only this time it's being produced by a completely different equation!

If many different equations produce similar chaos, then it may be that many different phenomena in the real world that appear chaotic—not just the way animal populations grow but the way the stock market performs and planets orbit the sun and the human brain works—follow the same rules underneath. So it's important to understand the bifurcation diagram. If we can understand why the bifurcation diagram looks the way it does, with areas of chaos interspersed with areas of order, we may understand a lot of things about how the universe works.

Mitchell Feigenbaum,
a physicist at Los Alamos
National Laboratory in
New Mexico and a leading
researcher in the quest to
understand chaotic phenomena

An important advance in understanding the bifurcation diagram was made only a few years after Robert May wrote his paper. A scientist named Mitchell Feigenbaum at the Los Alamos National Laboratory in New Mexico sat down with a calculator and studied the bifurcations produced by the logistic difference equation. He discovered that if you look at the distances between bifurcations in the diagram, these distances grow smaller by a constant ratio. The ratio is 4.669202609 . . . , where the dots mean that the number keeps going forever to the right of the decimal point. Not only does this ratio occur in the bifurcation diagram produced by the logistic difference equation but it also occurs in the bifurcation diagrams produced by all equations that produce bifurcation diagrams!

What is the significance of this magic number 4.669202609 . . . , now known as *Feigenbaum's number*? That's difficult to say, but the fact that it keeps popping up again and again in the various phenomena studied by chaos theory means that it must be an important number.

The bifurcation diagrams give us an important glimpse of chaos. What they show us is that simple equations and simple numbers don't always produce simple results. The logistic difference equation is simple enough, but the results that flow from it are so complicated that scientists do not yet fully understand them. Perhaps they never will. And if the universe can be described by a series of equations like the logistic difference equation, many of which produce equally complex results, then the universe itself could be a place of incredible complexity. Instead of a simple universe arising from simple equations, a very complicated universe may arise from simple equations.

This is one aspect of the phenomenon that scientists are now referring to as chaos. But it is not the only aspect. There is much more to chaos than just complicated events arising out of simple equations. Even as Robert May and Mitchell Feigenbaum were seeing one aspect of chaos in bifurcations, a number of other scientists were seeing other aspects in other places, including places as seemingly mundane and ordinary as the weather!

C H A P T E R

FIVE

Chaos in the Air

The idea that the universe can be described in terms of simple equations dates from the seventeenth century, when the great English physicist Isaac Newton used mathematics to describe the solar system. His description of the way gravity held the solar system together and kept it eternally in motion was so simple and mechanical that it made the solar system seem like a giant clock.

Like a clock, the planets and moons behaved in predictable ways. If you knew the numbers that described their motion and the equations that tied those numbers together, you could predict where each of the planets would be in a thousand, a million, even a billion years, just as you knew where the hands of a clock would be at midnight or at six in the morning.

In theory, you could predict what the planets would do. In practice, it wasn't quite so simple.

The equations that Newton used to describe the motions of the planets were a new type of equation invented by Newton himself. They were called *differential equations*. They were used to describe objects in constantly changing motion, the way that the planets were constantly changing direction as they moved in vast ellipses about the sun.

These differential equations could be used to predict the motions of the planets and other objects in the solar system with remarkable precision. But there was a catch. As long as you were dealing with only two objects, such as the sun and the Earth or the Earth and the moon, it was easy to solve the equations that described their motion and predict what they would be doing in the future. But the moment you added a third object to the equation—if you tried to predict the motion of the Earth, sun, and moon together, for instance—the equations became extremely difficult to solve. In fact, they became nearly impossible to solve.

This is known to physicists—scientists who study the fundamental ways in which objects, such as planets and atoms, behave—as the *three-body problem*. Understanding two bodies (or objects) is easy; understanding three bodies is a thousand times harder. And, as you add more bodies, the problem grows harder still. Alas, the solar system consists of much more than three bodies: it contains nine planets, one sun, and uncountable asteroids and comets. Predicting the future of the solar system with Newton's equations therefore turned out to be harder than it had at first seemed.

The problem is that all of these bodies are affecting all of the others. Every object in the solar system is producing gravity, a force that affects the behavior of other objects both large and small, and thus the way in which they interact is far from simple. It is, in fact, extremely complex. Scientists get around this problem by using somewhat simplified methods of solving Newton's equations. The answers they get are not precisely accurate, but they are very close to being accurate.

For roughly three centuries, scientists believed that this approximate accuracy was enough. Although the results that they obtained in predicting the future of the solar system may have contained errors, these errors would tend to cancel out one another. An error in one direction, for instance, would compensate for an error in the other direction. A scientist attempting to predict the future of the solar system could use techniques that were *approximately* accurate and get results that were *approximately* accurate. In most cases, that was good enough. That meant that the future of the solar system could be predicted.

And if the future of the solar system could be predicted, just about anything else could be predicted. Everything in the universe, including human beings, operates according to certain underlying rules. If we could understand what those rules are and state them as equations, we could predict the future by solving those equations.

This idea is generally credited to the eighteenth-century French mathematician Pierre Simon de la Place, who was born more than twenty years after Newton died. La Place imagined that a "demon" (by which he meant some imaginary force with unlimited powers of memory and mental calculation) could hold in its immense mind the position and velocity of every particle in the universe and use Newton's laws to predict where they would be at any time in the future. He wrote: "Such an intelligence would embrace in the same formula the movements of the greatest bodies of the universe and those of the lightest atom; for it, nothing would be uncertain and the future, as the past, would be present to its eyes." In other words, it could predict the future.

Today, we would speak not of a demon but of a giant computer. If we could tell a huge computer the exact position of every subatomic particle in the universe and program it with the equations that make the universe tick, the computer could tell us what would be happening a hundred years—or a thousand or a million or a billion years—from now. It could calculate the future as surely as our computer in the last chapter calculated the way that animal populations grow.

Of course, we ran into problems in the last chapter. The populations began to behave in a chaotic manner when certain variables were given certain values. But even this chaotic behavior wasn't random. It could be calculated. In fact, we *did* calculate it. We just have to know the right numbers to plug into the computer and it can calculate the rest.

But the moment we start thinking about a computer big

enough to calculate the future of the universe we run into additional problems. For instance, how big would a computer have to be in order to hold all the information about every subatomic particle in the universe? It would have to be as big as the universe itself, maybe even bigger. And that is a paradox. Even if the computer were no larger than the universe, it would still fill the entire universe—and thus would only be able to predict its own future, which is a pretty useless prediction. Perhaps we could locate our computer in another universe, but nobody knows whether universes other than our own even exist.

And how do you enter the exact position and velocity of a subatomic particle in a computer? How exact do these numbers have to be? When you put precise fractional numbers into a computer, you inevitably have to round them off a little. If you needed the numbers to be infinitely accurate, you would need an infinite amount of computer memory to store them in, because the numbers would have to be accurate to an infinite number of decimal places.

As a result of these two problems, our imaginary computer for predicting the future would have to be simplified somewhat. There's no way that we can enter the exact position of every particle in the universe into it. We'd have to restrict ourselves only to those particles in a certain portion of the universe—say, those particles that make up the planet Earth and its immediate surroundings—and we'd have to round off the numbers that describe them. That way, we could probably get by with a computer that isn't much larger than the solar

Pierre Simon de la Place
(1749–1827), French mathematician.
He believed in a mechanistic
universe in which Newton's
laws on motion could be
used to predict future phenomena.

system. And we could locate it in a neighboring solar system. Although this would be a pretty spectacular engineering task, it's not theoretically impossible.

Would such a computer be able to predict accurately the future of the planet Earth? Probably not—for reasons that we will see in a moment.

Although nobody has seriously proposed building a computer large enough to predict the future of the universe, or even of the planet Earth, it *has* been suggested that we build computers that can predict the future of some small part of what is happening here on Earth. In fact, it's already been done. For instance, there are several large computers dedicated to predicting the future of the weather.

Weather, loosely defined, is the sum of all the things that happen in the lower part of the Earth's atmosphere, the part where human beings live. These phenomena, which are all caused by the way that the sun's heat moves through the atmosphere, include rain, snow, wind, changes in temperature, and a number of other variables. These factors are very important to human beings because they directly affect our lives. It is convenient to know in advance when it is going to rain or snow, so that we can plan accordingly. And it is crucial that we know when there is going to be a devastating storm such as a hurricane or tornado, so that we can save the lives and property that would otherwise be lost in the storm.

Of course, there have been weather predictions available for many years. All we have to do is pick up the phone and dial a special number or turn on the television to watch the news and we can get a weather prediction.

But these weather predictions are far from perfectly accurate. In fact, they aren't even couched in terms of perfect accuracy. They are stated in *probabilistic* terms. The weather forecaster tells us that there is a 50 percent *probability* of rain, which means that there is also a 50 percent probability of no rain.

Wouldn't it be nice if we could have weather predictions couched in *deterministic* terms: that is, predictions that tell us it *will* rain or it *will* be clear? And wouldn't it be nice if these predictions could always be correct?

When digital computers came along in the 1940s, it seemed that this was a genuine possibility. The equations that described the weather were well known to meteorologists, the scientists who study weather phenomena. If they could program these equations into a fast enough computer and plug in the numbers describing today's weather, the computer could calculate *exactly* what the weather was going to be tomorrow. There would be no need for probabilistic forecasts. The weather forecasts could be deterministic—and perfect.

The problem, initially, was that there weren't any computers even remotely fast enough to do this. And there was no way to gather enough information to tell the computer about the state of today's weather even if the computer were capable of absorbing all of that information.

But this situation has changed with time. Today there are computers million of times faster than the first computers available in the 1940s and there are weather satellites and other sophisticated information-gathering devices that can give us volumes of information about the current state of the weather. Of course, it's still not possible to know *everything* about the current state of the weather—the position of every particle of air or the velocity of every breeze—but as we said earlier in this chapter, it should be possible to use approximate information and still get good results. There would be errors in the data that we give to the computer, sure, but the errors should cancel one another.

And yet it's still not possible to make perfect weather predictions. What went wrong?

Chaos is what went wrong. In 1961, an American meteorologist named Edward Lorenz actually programmed a model of the weather into a computer. It wasn't the real weather that Lorenz programmed into his computer. It was imaginary weather. It was weather on a planet that didn't really exist. But the equations that Lorenz used to describe the way this imag-

inary weather acted were close approximations of some of the equations that describe real weather. Lorenz hoped that by studying the imaginary weather in his computer, he could learn important things about the way weather behaved in the real world. And, indeed, he did.

Lorenz's imaginary weather really behaved like the real weather. It had clouds and storms, cold fronts and warm fronts. All of these took the form of numbers that were typed out on a printer attached to the computer. But Lorenz knew that these numbers represented real weather phenomena taking place in the imaginary weather universe inside the computer.

One day, Lorenz was examining the printout from a previous day's simulated weather. He decided that he wanted to study that particular weather again. Unfortunately, Lorenz had no way of returning to that day's imaginary weather except by typing the numbers on the printout back into the computer. So he entered the numbers into the computer one by one. But he didn't type them exactly as they had been stored in the computer on the previous day. Lorenz's computer program performed arithmetic on numbers that could have up to six digits to the right of the decimal point, like this: 1.482603. But the printout that Lorenz was using showed only the first three digits after the decimal point.

Lorenz assumed that this wouldn't matter. Tiny errors, such as the missing digits to the right of the decimal point, should cancel one another out. And errors that begin on the

Edward Lorenz of the
Massachusetts Institute of
Technology (MIT).
Although trained as a
meteorologist, Lorenz did
research on weather
forecasting which led him to
astonishing mathematical insights.

fourth digit after the decimal point are very small errors indeed.

And yet when Lorenz restarted his simulated weather using the numbers that he had typed in from the printout, the computer produced weather that was different from the weather the first time he had run this particular simulation. At first, the differences were small. But, as he continued to run the simulation, the differences became larger—until the weather on the computer was completely different from what it had been the day before.

How could that be? He was using the same numbers as the day before, albeit rounded off after the third decimal place. Surely it wasn't possible that such tiny errors could create completely different weather!

But they did. And it didn't take Lorenz long to figure out what he was seeing. He was seeing evidence that perfect weather forecasts over a period longer than a couple of days were impossible.

Lorenz had discovered evidence of something that meteorologists had long suspected. They even had a name for their suspicion. They called it the *butterfly effect*. The name was based on the half-whimsical belief that a butterfly flapping its wings in Asia could affect the weather in New York a few days or weeks later. In other words, small causes could be magnified over a period of time to have big effects.

The imaginary weather on Lorenz's computer was demonstrating the truth of that belief. The tiny errors that were created when Lorenz typed in the numbers without the last three decimal places were like the tiny effect that a butterfly beating its wings has on the atmosphere. And those errors were having a huge effect on the imaginary weather.

This meant that perfect weather predictions would never be possible because there will always be tiny mistakes in our information about the weather. No matter how many weather satellites and weather balloons and wind socks and thermometers a weather forecaster uses to gather data, there will always be tiny pieces of information that will be missed, the equivalent of a butterfly flapping its wings in a location where it can't be detected by weather satellites and wind socks. And even if we could have perfect knowledge of the weather, it would be impossible to put that information into a computer without rounding off the numbers to a certain number of decimal places. Lorenz's results proved that this wouldn't work.

Scientists have another name for the butterfly effect. They call it *sensitive dependence on initial conditions.* What this means is that the tiny differences in two sets of data don't cancel each other. Instead, some small differences become big differences and others just fade away. Thus, tiny errors in data can grow up to be big errors. Since errors are inevitable, perfect predictions of any system that displays sensitive dependence on initial conditions are impossible. And weather, according to Lorenz, is a system that shows sensitive dependence on initial conditions.

Like Robert May more than a decade later, Edward Lorenz had discovered chaos.

But just as the chaos in the logistic difference equation contains a strange kind of order, so does the chaos that Lorenz discovered in the weather. And just as that order became obvious when we graphed the logistic difference equation, so Lorenz saw the order in the weather when he drew a picture of it.

The graph of the weather that Lorenz created was based on three equations that described a phenomenon known as *convection,* which is how heat moves through air. Since weather phenomena are caused by the movement of the sun's heat through the air, convection is of central importance in understanding the weather.

When Lorenz finished graphing these equations, they produced a very odd picture indeed. (See Figure 5.1.) Although the convection equations demonstrated the same unpredictable "sensitive dependence on initial conditions" that Lorenz's weather program had, the graph produced by these equations had a strange predictability. It consisted of a line that described a kind of twisted figure eight, almost butterflylike in shape. In

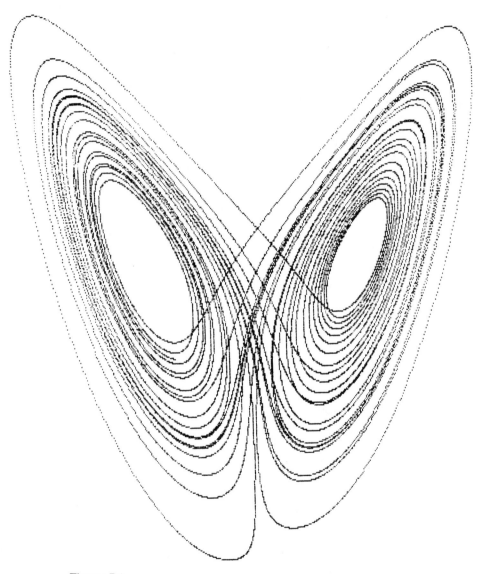

Figure 5.1.
Order within disorder: the Lorenz attractor
(Courtesy of Edward Lorenz)

fact, this graph has since become known as *Lorenz's butterfly*—no relation to the butterfly effect. What was really odd about this graph, though, was that it never repeated itself. Despite the fact that the line keeps describing almost the same shape over and over again, it never describes *exactly* the same shape. Even if Lorenz had kept graphing this equation for the rest of eternity, it would never have repeated itself. In fact, the line in the graph never even crosses itself. (In the picture, it *seems* to cross itself. But that's because Lorenz's graph is actually a *three-dimensional graph*. It has three axes instead of two, so that the line is actually moving in three-dimensional space. Because you are seeing it projected onto a two-dimensional piece of paper, you can't see that the line actually passes in front of itself in places where it seems to cross itself.)

In the last chapter, we saw how certain equations, when they were iterated enough times, would start producing the same results over and over again. We called this repeating result an attractor. Lorenz's butterfly is also a kind of attractor, because it repeats the same shape over and over again. But it is not an ordinary attractor, because it never repeats itself precisely. Instead, it is a *strange attractor*.

Lorenz realized that he had discovered something important indeed. Perhaps, if the underlying equations that described the weather behaved in this oddly repetitive way, it might be possible to predict them after all, despite the sensitive dependence on initial conditions. If only the strange attractor could be understood, it might tell scientists things about the weather that they had never understood before.

But when he published a paper about his strange attractor in a meteorological journal, it was largely ignored. Nobody was interested in hearing about strange attractors or sensitive dependence on initial conditions.

Not, that is, until a little over a decade later, when strange attractors started turning up all over the place.

SIX

Chaos Everywhere

In the 1970s, scientists in a number of fields became aware that the phenomena that they were studying behaved in a chaotic manner. These phenomena displayed complicated behavior despite the fact that they could be described with simple equations. And they had a sensitive dependence on initial conditions.

They had a strange predictability about them too. And the key to that predictability was the strange attractor. When the phenomena that these scientists were studying were reduced to equations, and those equations were graphed, they produced strange attractors.

The strangest thing about strange attractors is that they have a fractional number of dimensions. What's that again? Well, everybody knows that we live in a three-dimensional world. The three dimensions are length, breadth, and depth. A straight line has one of these dimensions, a sheet of paper (or a *plane*, as a mathematician would call it) has two—and space has three.

But mathematicians also regard certain objects as having dimensions part of the way between these dimensions: that is, they have fractional dimensions, such as 1.3 dimensions or 2.6 dimensions. Take, for instance, a jagged line. Although straight lines ordinarily have one dimension, a jagged line also extends partially into a second dimension. Yet it isn't quite a plane either, so it doesn't have a full two dimensions. It has a fractional dimension partway between one and two dimensions. Similarly, a crumpled piece of paper would have something between two and three dimensions.

Mathematicians even have methods of measuring how many dimensions an object has. They can use these methods to tell you that a particular jagged line, such as the jagged coastline of a continent, has (say) 1.13 dimensions. For example, the coastline of Britain has a fractional dimension of 1.26.

What good does this information do us? Strange attractors, like the one that Lorenz discovered in his weather equations, have fractional dimensions. And different strange attractors produced by different equations have different fractional dimensions.

Scientists aren't quite sure yet what this means. But it may turn out to be possible to make predictions about chaotic systems based on the dimensions of their strange attractors alone. In fact, there's at least one researcher who believes he can use the dimensions of a strange attractor to read people's minds! (Other scientists disagree with him, however.)

Unfortunately, scientists have also encountered a problem in identifying chaos. The problem is that they haven't quite defined chaos yet. Roughly speaking, scientists know chaos when they see it. Proving that chaos exists in a certain system, though, is easier said than done.

Basically, scientists say that chaos exists in a system when it shows a sensitive dependence on initial conditions: that is, when tiny changes in a system turn into major changes over a period of time—and when a strange attractor shows up when the variables in the system are graphed.

And these signs of chaos are appearing all over the place. Here are some of the areas where scientists have encountered chaos in recent years:

ASTRONOMY

Ironically, one place that chaos has turned up is in the very place that Newton found the pure mathematical order that convinced scientists that the universe was basically simple. That place is the planets of the solar system.

As we saw in the last chapter, scientists studying the solar system after Newton were forced to use simplified versions of his equations because the mathematics involved in calculating the positions of the planets and their moons was too complicated to be solved. But the arrival of the digital computer changed that situation somewhat. Computers can often perform difficult calculations when human beings can't. It's not that computers are smarter than human beings, but that they don't get bored with performing iterated equations over and over again. And complicated equations don't intimidate them; they just plunge in and solve them, no matter how much work is involved.

As a result, modern scientists have been able to explore the equations that describe the motions of the planets in more detail than Newton could.

One scientist who has studied the solar system in some detail is Jack Wisdom of the Massachusetts Institute of Technology. Wisdom's first area of study was the asteroid belt, an area filled with large chunks of rocky debris in orbit around the sun. Some scientists believe that the asteroid belt is the remains of a planet that didn't fully form when the solar system was being born. They also believe that the asteroid belt may be the source of many of the meteorites that bombard the Earth.

But what would cause an asteroid to jump out of the asteroid belt and fling itself in the direction of Earth to become a meteorite? This was the question that Wisdom asked himself. And as he studied the behavior of the asteroids he learned the answer.

The next planet beyond the asteroid belt is the giant planet Jupiter, largest in the solar system. Jupiter is so large that its gravity has a strong effect on the orbits of the asteroids. Remember, in Chapter 4, how certain values of the variable R

caused the logistic difference equation to produce chaotic results? Well, Wisdom discovered that when asteroids are a certain distance from Jupiter, the giant planet's gravity (which we can think of as a variable in Newton's equations) takes on just the right value to cause the asteroids to act chaotically. Suddenly they begin to move in strange patterns. They can even shoot right out of the asteroid belt, toward the sun. And that's why they sometimes become meteorites bombarding the planet Earth.

This also explains why astronomers studying the asteroid belt have noticed that there are "gaps" in it, regions where there are either no or very few asteroids. These gaps represent the "chaotic belt," the area where Jupiter's gravity takes on just the right value to make the asteroids behave chaotically. And eventually these chaotic asteroids either move back out of this chaotic region or shoot off toward other parts of the solar system, which is why there are almost no asteroids within this region.

Wisdom then turned his attention to Hyperion, one of the moons of Saturn. Recent space probes have shown that Hyperion describes a weird, tumbling orbit about the ringed planet. Sure enough, when Wisdom began studying it, he found evidence of chaos in Hyperion's orbit.[1]

In order to look for chaos elsewhere in the solar system, Wisdom teamed up with the MIT computer scientist Gerald Sussman to build a special computer for plotting the motions of the planets. The computer that they built is known as the digital orrery. (*Orrery* is a word that came into use in the seventeenth century to describe mechanical models of the solar system.) The digital orrery is not a general-purpose computer. In fact, it can't do anything *but* plot the orbits of planets. But for that one purpose, it is extremely fast and accurate.

The digital orrery allowed Wisdom and Sussman to plot the history of the solar system for 845 million years into the future. During that period, many of the planets displayed certain peculiarities in their orbits, but the one that was the most obviously chaotic was Pluto.

The planet Pluto is the outermost planet of the solar system, but it glides about the sun in an orbit that is so lopsided that for the final years of the twentieth century it will actually be closer to the sun than the next innermost planet, Neptune, making Pluto temporarily the eighth planet instead of the ninth.

According to Wisdom and Sussman, Pluto's orbit displays all the signs of chaos. They suspect, however, that only part of the orbit is chaotic. For the moment, Pluto may be in a chaotic zone much like that occupied by the asteroids Wisdom believes have been hurled out of the asteroid belt toward the sun. Could Pluto one day be flung out of its orbit into the inner solar system or the depths of space? It's possible, but Wisdom and Sussman can't say for sure.[2] It may even be that Pluto once had a completely different orbit around the sun than it does now but was shifted chaotically into its present orbit at some time in the distant past.

As Wisdom and Sussman have concentrated primarily on the large outer planets of the solar system, J. Laskar of the Bureau des Longitudes in Paris has been studying the inner planets and believes that he has found chaos in their orbits too. Even the orbit of our own planet, Earth, may be chaotic. It's unlikely that Earth will ever shoot out of its orbit—at least not in the foreseeable future—but Laskar's findings indicate that it may be impossible to predict the future orbital motion of our planet beyond a certain distance in the future, just as Edward Lorenz found that the weather could not be predicted beyond a certain distance in the future.[3]

EPIDEMIOLOGY

For many years, epidemiologists, scientists, and doctors who study the way that epidemic diseases spread through the population, have been seeking a way to predict the future progress of epidemics. If there were some way to predict how severe an epidemic was going to be—or, even better, when an epidemic was going to start—it would be possible for doctors to intervene at appropriate moments with vaccinations and medicine to prevent massive death and illness.

To this end, they have developed a number of equations that describe the way epidemics spread. But matching these equations to the real world has proved difficult. Not surprisingly, they have tended to blame these difficulties on noise, unwanted variables that complicate the way that epidemics spread in the real world, as opposed to on paper or in the computer.

Epidemics don't spread randomly. There are rules to the way that they move through a population. For instance, an epidemic of a contagious disease will begin to die off when enough people in a population are immune to the disease so that carriers of the disease are no longer likely to encounter anyone to whom they can transmit it. Calculating when this event occurs is an important part of calculating the course of an epidemic.

Although they are not random, graphs of the way that epidemics spread show a strange combination of randomness mixed with order. Epidemics occur on seemingly regular schedules. Yet the precise timing and the severity of the epidemics are not always predictable. Is this randomness the product of noise? Or is it chaos?

Epidemiologists with an interest in chaos theory are now beginning to suspect that epidemic diseases follow a chaotic pattern as they spread through the population. The epidemiologist William Schaffer of the University of Arizona has even developed equations that produce chaotic behavior that he believes is very close to the way that epidemics behave in real life.

Could these new equations offer a method of predicting the course of epidemics that is more accurate than the nonchaotic equations used by earlier epidemiologists? Schaffer thinks so, but other epidemiologists are not so sure. Even if Schaffer is correct, this doesn't necessarily mean that doctors will know exactly when every epidemic will arrive and will have the appropriate treatment ready when it does. But it may help epidemiologists to understand the behavior of epidemics much better than they do at present.[4]

THE HUMAN BODY

The human body is often compared to a machine, because it is made up of a large number of parts that work together in precise coordination. One of the most important of those parts is the human heart, which pumps the blood that carries oxygen to the individual cells that make up the body. The heart beats with a rhythm that can be measured by an *electrocardiogram* (*ECG* for short). By studying this rhythm, doctors can sometimes tell whether an individual's heart is working correctly or whether that individual may be headed for a heart attack.

In order to determine whether the heart is producing the correct rhythm, it's necessary first to know just what the correct rhythm is. In the past, doctors have assumed that a healthy heart beats with a steady, constant rhythm. Any sudden changes or peculiar patterns in the heart's rhythm must be a sign of bad health.

Now, however, a few researchers are suggesting that an unpredictable, chaotic rhythm may be a sign of a healthy heart, and that a steady, predictable rhythm may be a warning of heart disease! In other words, chaos in the heart is healthy.

In an effort to demonstrate chaos in the heart, researchers have taken cells from a chicken heart, cultured them in the laboratory, and exposed them to an electric current. When the flow of the current is timed correctly, the cells begin to behave chaotically. Could this chaos be a property of the entire heart as well as its individual cells?

Ary Goldberger of the Harvard Medical School thinks so. He has analyzed the way that the human heartbeat changes from moment to moment and has concluded that the changes represent a chaotic pattern. And when he compares the moment-by-moment variations in the heartbeat of a healthy person to those of the heartbeat of a person suffering from heart disease, the healthy heartbeat is actually more chaotic than the unhealthy heartbeat.

Other researchers in this area have been cool to Goldberger's results, believing that the chaotic nature of the heart isn't relevant to cardiac medicine. But Goldberger believes that his insight may lead to new methods of treating heart diseases (not to mention new methods of maintaining cardiac health) in the future.

A few scientists who study the brain have become interested in chaos. Walter Freeman, a biologist at the University of California at Berkeley, has studied the electrical signals generated by the brains of rabbits and believes that they create chaotic patterns. Specifically, he observed the parts of the rabbits' brains that process information about odors and discovered that the chaotic patterns produced by these parts of the brain change according to the odor that the rabbit is smelling.[5]

Paul Rapp, a brain researcher at the Medical College of Pennsylvania, even believes that he can tell what people are thinking by measuring the fractional dimensions of their brain waves. By using a computer to graph the signals produced by the human brain, Rapp has discovered patterns that resemble strange attractors and has measured their fractional dimensions. He says that these dimensions change depending on what the individuals producing the signal are thinking about. For instance, a person thinking about a simple problem may produce a mental strange attractor with a fractional dimension of 2.3 and a person thinking about a tougher problem may produce a strange attractor with a higher dimension, around 2.9. In effect, Rapp can use these strange attractors to read peoples' minds!

Both Rapp and Freeman believe that the chaotic activity of the brain is essential to creative thought. Because chaotic systems can change from one state to another very rapidly, the chaos in our brains allows us to cast about from one idea to another in a hurry. It is this chaos that allows us to have new ideas, "inspirations." Freeman even suggests that this chaos may be responsible for consciousness itself, the mysterious ability of the brain to be aware of its own thoughts. It's important to note, however, that these ideas about chaos in the brain are quite controversial. There are many scientists who consider them premature, at best.[6-8]

METEOROLOGY

Edward Lorenz first saw chaos in a set of equations describing the way convection carries heat through air, equations that are central to the way that weather works. But is there evidence that the chaos Lorenz glimpsed in his equations actually exists in the real weather?

Oddly, much of the research on chaos in the weather has not been focused on the weather here on planet Earth. Instead, it has focused on the weather on the planet Jupiter.

Why Jupiter? Ever since astronomers first looked at Jupiter through telescopes in the seventeenth century, they have been aware of a strange spot on the surface of the planet. As improved telescopes have made it possible to see the color of this spot, it has become known as the Great Red Spot of Jupiter. It is the giant planet's most prominent feature.

But what *is* the Great Red Spot? According to twentieth-century astronomers, the "surface" of Jupiter that we see through telescopes isn't really the planet's surface. Rather, it is a thick layer of clouds covering the planet's actual surface. (In fact, Jupiter may not even have a surface in the sense that the inner planets like Earth and Mars have surfaces.) The Great Red Spot, therefore, must be some kind of weather phenomenon in the clouds of Jupiter, perhaps a giant hurricane.

Yet if the red spot is a giant hurricane, it is a hurricane that

Could the Great Red Spot in Jupiter be a giant strange attractor in the planet's atmosphere?

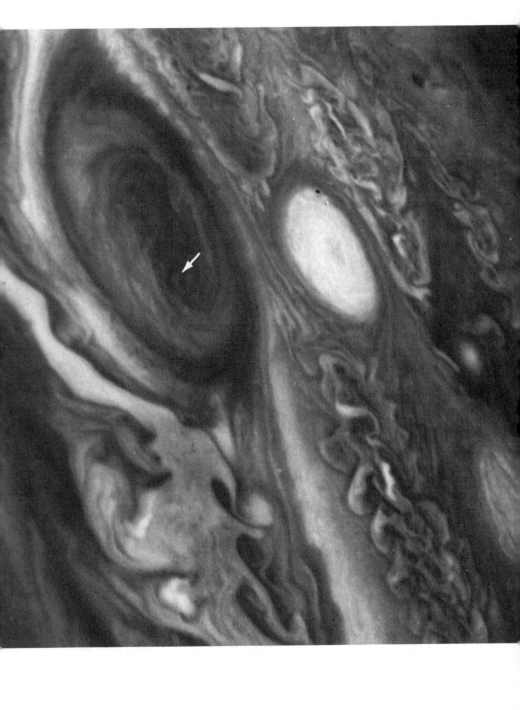

has been swirling for at least three hundred years, and maybe a great deal longer. Hurricanes on Earth come and go in a few weeks at most. How could a hurricane last that long?

Chaos theorists, however, believe that the red spot is some kind of chaotic phenomenon, a kind of strange attractor in the sky. Indeed, Philip Marcus of the University of California at Berkeley has designed a mathematical model of the Jovian atmosphere and has found that a spinning vortex of clouds is created naturally by his equations. Although his model is quite simple, Marcus believes that it may actually explain the existence of the red spot.

As for chaos in Earth's weather, some theorists believe they have found strange attractors in this planet's climate. And Anastasios Tsonis of the University of Wisconsin—Milwaukee believes that these strange attractors can be used to predict the weather. He proposes that scientists analyze the weather and determine what sort of strange attractor it produces, then use the strange attractor in turn to determine where the weather is going next. He believes that this may provide a simpler and more accurate method of predicting the weather than more conventional techniques. Other meteorologists disagree, however.[9]

These are only a few of the areas where chaos has turned up. There are others.

Turbulence Turbulence, in fact, is one of the first areas where chaos was observed. Turbulence is what happens to air or water or any other smoothly flowing stream of particles when its smooth flow is disturbed. Scientists have had trouble understanding turbulence for centuries, because the equations that describe its behavior are extremely complicated. And this creates problems in everything from the design of airplane wings (which must be curved so that air flows past them smoothly) to the building of water pipes. Chaos theory may offer insights into turbulence and ways of preventing it.

The Stock Market Financial experts have been looking for ways to predict the actions of the stock market for years, but nothing seems to work. Now it has been suggested that the ups and downs of the stock market may follow some kind of strange attractor. It is possible (though perhaps not probable) that chaos theory may lead to a method of understanding the stock market and predicting its twists and turns.

When chaos turns up, it's important that scientists know how to recognize it. In the next chapter, we'll take a look at the face of chaos—and visit a strange world that is made up of chaos itself.

C H A P T E R
SEVEN

Pictures of Chaos

It's impossible to talk about chaos without talking about *fractals*.

More than thirty years ago, a French-born mathematician named Benoit Mandelbrot began to notice strange patterns in the world about him. In particular, he began to notice that nature had a penchant for repeating itself, but in a peculiar way.

Consider a tree. It is made up of a trunk, with large branches sticking out from it. These branches have smaller branches sticking out of them, which in turn have twigs sticking out of them. And the twigs have even smaller twiglets sticking out of them. Yet at every level—trunk, big branch, little branch, twig, twiglet—the tree looks pretty much alike. The branches look like the trunk and the smaller branches look like the bigger branches.

The tree, we might say, is *self-similar:* The parts look like the whole.

Mandelbrot noticed that this was also true of economic phenomena that he was studying. When he looked at the way the stock market rose and fell from day to day and graphed it on a sheet of paper, it looked pretty much like the way the stock market rose and fell from year to year, only on a smaller scale. The numbers were bigger in one case than the other, but the rising and falling patterns were the same. The rising and falling patterns are self-similar. The parts look like the whole.

Or consider the coastline of a continent. When viewed from outer space, it looks jagged. When viewed from an airplane, smaller details are visible, but they look jagged in exactly the same way. And when viewed by a person standing next to the shore, even more details can be seen—and those details still look jagged. Even when observed through a magnifying glass, the coastline appears jagged in much the same way as it does from the satellite, though it's being viewed at a much lower level of detail. The coastline is self-similar: The parts look like the whole.

Mandelbrot concluded that this self-similarity must be an important part of nature. He decided that it needed a name. He wanted a name that reflected the fragmented, fractured nature of coastlines, trees, and stock market graphs. He also wanted a name that reflected the fact that these objects also tended to have fractional dimensions. He took the Latin word *fractus*, which means "to break," and changed it slightly to produce the word *fractal*.

Fractals, according to Mandelbrot, are things that have the quality of self-similarity. Trees are fractal. The stock market is fractal.

Even the human body is fractal. The bloodstream, for instance, is carried through a network of veins that branch much as the limbs of a tree do. As the blood passes into smaller and smaller veins branching off larger and larger veins, the oxygen in the blood can eventually reach even the most remote cells of the body. The human nervous system and lungs are constructed in a similar, fractal manner.

Chaotic systems also tend to be fractal. A strange attractor, for instance, contains patterns that look the same at every level. The parts look like the whole. A strange attractor, therefore, is fractal.

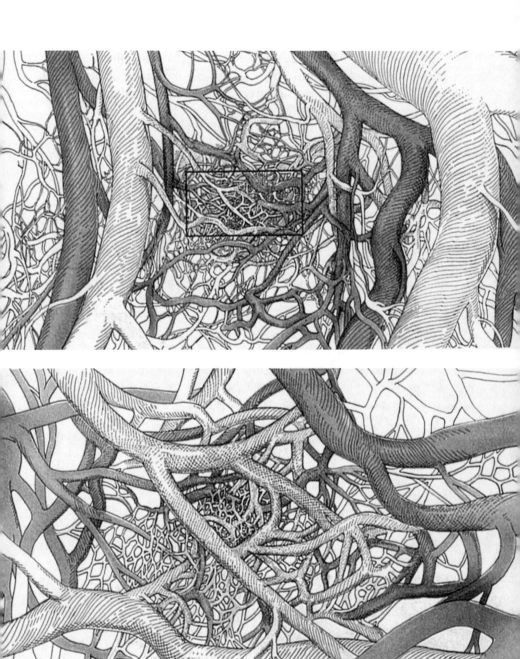

The bifurcation diagram that we studied in Chapter 4 is fractal. Remember how images of the entire diagram kept appearing as tiny details within the whole diagram? The parts resemble the whole.

Chaos theorists have devised systems of fractal mathematics that can be used to describe fractal structures. In conjunction with the computer, they can even be used to reproduce such structures. Moviemakers are now using fractal techniques to create computerized images of fractal objects such as trees and mountain ranges. In the movie *Star Trek 2: The Wrath of Khan,* for instance, the image of a raging fire sweeping across the surface of an imaginary planet was created with the aid of fractal mathematics.

But perhaps the most astonishing byproduct of Mandelbrot's research into fractals is the *Mandelbrot set,* a mathematical structure that can be used to create startlingly beautiful images of a chaotic, fractal world that only exists in the abstract world of numbers. In the rest of this chapter, we'll show you how to use a computer to explore the Mandelbrot set, as an astronaut explores a brand new planet, but, in this case, a planet that is infinitely large!

What is the Mandelbrot set? In mathematics, a set is a group of things that all have something in common. The readers of this book, for instance, belong to the set of human

The new sciences of complexity have applications to biomedical research. The boxed area in the upper drawing of blood vessels is seen in an enlarged view in the lower drawing. The vessels display fractal-like branching.

beings and to the set of book readers. If you happen to be male, then you also belong to the set of male human beings. And if you happen to be female, you belong to the set of female human beings. (It's possible for something to belong to more than one set. In fact, most things do.)

Numbers belong to various sets. For instance, the numbers 2, 4, and 8 belong to the set of even numbers. The numbers 3, 6, and 9 belong to the set of numbers divisible by 3. The numbers 2, 3, 5 and 7 belong to the set of prime numbers. And so on.

The Mandelbrot set is made up of numbers. But they are a special kind of number, called *complex numbers*. If you haven't studied complex numbers yet, you don't really need to know what they are in order to understand the Mandelbrot set. All you need to know is that complex numbers are made up of two parts, called the *real* part and the *imaginary* part. (These are just names. The "imaginary" part is every bit as real as the "real" part, despite the way that the names make them sound.) A typical complex number might look like this: $14 + 3i$. The first part (14) is the real part. The second part ($3i$) is the imaginary part. (The i simply marks the second part as imaginary.)

Just like ordinary numbers (which are called *real numbers* by mathematicians), complex numbers can be graphed. In fact, it's difficult to envision complex numbers without graphing them. We tend to imagine real numbers as existing on a kind of line: a *number line*. (See Figure 7.1.) Zero is in the middle of the line, positive numbers stretch out to infinity to the right of zero, and negative numbers stretch out to infinity to the left of zero.

Similarly, complex numbers exist on a plane, the *complex plane*. The complex plane is actually a kind of graph, where the x axis is a horizontal number line with the real numbers on it and the y axis is a vertical number line with the imaginary numbers on it. (The positive imaginary numbers stretch out to infinity above zero and the negative imaginary numbers stretch

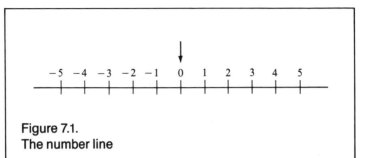

Figure 7.1.
The number line

out to infinity below zero.) As in any graph, the two lines cross at zero.

To find any complex number on this plane, you simply find the position of the real part of the number on the x axis and the position of the imaginary part on the y axis and graph them: that is, find the point where they intersect. That's the location of the complex number. For instance, to find the complex number $-4 + 3i$ on the complex plane you would find -4 on the x axis and 3 on the y axis. The point where they intersect represents the complex number $-4 + 3i$.

Complex numbers can be fractional, just like real numbers. In fact, most of the complex plane is made up of fractional numbers, such as $1.23432 + 7.232112i$. Of course, the complex plane is infinitely large, so there is an infinite quantity of both whole numbers and fractional numbers on the complex plane. But there is also an infinite number of fractional numbers between every two whole numbers. In fact, there is an infinite number of fractional numbers between any two *fractional numbers!* Thus, we can say that any part of the complex plane, no matter how small it may seem, is infinitely large, because we can always zoom in to examine the fractional numbers within it.

Benoit Mandelbrot discovered that when you perform a certain series of mathematical operations (which we will not describe further) on numbers in the complex plane, they respond in one of two ways: Either they become very big, growing rapidly toward infinity, or they stay very small. The numbers that stay very small when these operations are performed on them are said to be in the Mandelbrot set. The numbers that grow rapidly toward infinity are said not to belong to the Mandelbrot set.

Mandelbrot tried graphing this set of numbers on a piece of paper, by putting black dots on the complex plane to represent numbers that belong to the Mandelbrot set. When he did, the graph formed an interesting pattern. It looked almost like a large bug. When he used a computer to graph the set, he found that he could "zoom in" on smaller and smaller fractional parts of the set and see more and more details. What he saw when he zoomed in were more strange patterns, many of which also looked like bugs—that is, they were self-similar images of the entire Mandelbrot set. The Mandelbrot set was fractal.

In the years since Mandelbrot first graphed the set that bears his name, many mathematicians, computer programmers, and amateur fractal aficionadoes have also graphed the Mandelbrot set, mostly by using a computer to do the graphing. The images of the set that they have produced are astonishing. Many of these graphs add color to the images by painting the dots that represent the numbers outside the Mandelbrot set in different hues, depending on how quickly they grow toward infinity when Mandelbrot's operations are performed on them.

The astonishing thing about the Mandelbrot set is the level of detail inside it. No matter how far you zoom in on this fabulous Mandelbrot world, there is always new detail inside it—and some of the details are surprising indeed. There are many self-similar images of the entire Mandelbrot set within the set, yet none of them is identical to the complete image of the set. They all have differences, sometimes large, sometimes small. And new features appear at deeper levels of details, such as patterns that resemble seahorses and filigreed jewelry.

If you have a computer that can display graphics, one of the most endlessly fascinating things that you can do with it (short of playing video games) is to explore the Mandelbrot set. So we've included a computer program that will allow you to explore the heart of the Mandelbrot set with an IBM-compatible computer. Instructions for typing the program and adapting it for various graphics adapters follow the program. Depending on which graphics adapter and resolution you are using, you will see as many as 256 different colors or as few as 4.

Whether or not you choose to type the program on a computer, we'll spend the rest of this chapter exploring the Mandelbrot set. Ready? Then let's go!

MANDELBROT PROGRAM

Here's the program for graphing the Mandelbrot set:

```
10   DEFSNG A-Z
20   SCREEN 1: KEY OFF
30   MAXDWELL = 150
40   NUMCOLORS = 4
50   NUMROWS = 100
60   NUMCOLS = 100
70   YOFFSET =1
80   XOFFSET = 1
90   INPUT "Lower lefthand corner, real part"; ACORNER
100  INPUT "Lower lefthand corner, imag. part"; BCORNER
110  INPUT "Length of side"; SIDE
120  CLS
130  COLOR 1
140  LINE (0, 0)−(NUMCOLS + XOFFSET, 0).
150  LINE (NUMCOLS + XOFFSET, 0) − (NUMCOLS +
     XOFFSET, NUMROWS + YOFFSET)
160  LINE (NUMCOLS + XOFFSET, NUMROWS +
     YOFFSET)−(0, NUMROWS + YOFFSET)
170  LINE (0, NUMROWS + YOFFSET)−(0, 0)
180  LOCATE 17, 1
190  PRINT "Percentage complete = 0"
200  PRINT "Dwell for last pixel = 0"
210  PRINT "Largest dwell = 0"
215  PRINT "MaxDwell="; MAXDWELL
```

```
220  PRINT "Real part = "; ACORNER
230  PRINT "Imaginary part ="; BCORNER
240  PRINT "SIDE ="; SIDE
250  HIGHDWELL = 0
260  GAP = SIDE / NUMROWS
270  AC = ACORNER
280  FOR X = XOFFSET TO NUMROWS − 1 + XOFFSET
290  AC = AC + GAP
300  BC = BCORNER
310  FOR Y = YOFFSET TO NUMCOLS − 1 + XOFFSET
320  BC = BC + GAP
330  AZ = 0
340  BZ = 0
350  COUNT% = 0
360  SIZE = 0
370  WHILE (SIZE < 4) AND (COUNT% < MAXDWELL)
380  TEMP = AZ * AZ − BZ * BZ + AC
390  BZ = 2 * AZ * BZ + BC
400  AZ = TEMP
410  SIZE = AZ * AZ + BZ * BZ
420  COUNT% = COUNT% + 1
430  WEND
440  COLOR 1
450  LOCATE 18, 23
460  PRINT COUNT%; " ";
470  IF (COUNT% < MAXDWELL) AND (COUNT% >
     HIGHDWELL) THEN HIGHDWELL = COUNT%:
     LOCATE 19, 16: PRINT HIGHDWELL
480  IF COUNT% = MAXDWELL THEN PSET (X,
     NUMROWS − Y + 1), 0 ELSE PSET (X, NUMROWS
     − Y + 1), COUNT% MOD (NUMCOLORS − 1) + 1
490  NEXT Y
500  LOCATE 17, 22
510  PRINT 100 * X / NUMCOLS; " ";
520  NEXT X
530  AS = INPUT$(1)
```

This program will draw the Mandelbrot set, or portions of the Mandelbrot set, in the upper left-hand corner of your computer's video screen. It doesn't use the entire screen because that would take too long to draw.

As written, the program will work on any IBM-compatible computer with a color graphics adapter. If you have the IBM

CGA adapter or an equivalent, you should leave the program exactly as it is. If you have the higher-resolution EGA or VGA adapter, you can make some simple changes so that the program will show more colors and/or more detail (though the program will work as written on those adapters).

If you have the EGA adapter, you should make the following change in order to see more colors:

```
20   SCREEN 7:KEY OFF
40   NUMCOLORS=16
```

The program will now display sixteen different colors instead of the four colors of the CGA version. And if you have the VGA graphics adapter, make these changes:

```
20   SCREEN 13:KEY OFF
40   NUMCOLORS=256
```

The program will now display 256 different colors.

You can also increase the *resolution* of the program—that is, the amount of detail that it shows on your screen. This is actually a more interesting change than increasing the number of colors. But it will also slow the program. Drawing high-resolution pictures will take approximately four times as long as drawing low-resolution pictures, and the low-resolution pictures are already pretty slow. Still, if you have a fairly fast computer you might want to give these changes a try. (Note also that VGA adapters can only display 16 colors in the high-resolution version, rather than 256 colors.)

If you have an EGA graphics adapter, make these changes:

```
20   SCREEN 9:KEY OFF
40   NUMCOLORS=16
50   NUMROWS=200
60   NUMCOLS=200
```

If you have the VGA graphics adapter, make these changes:

```
20   SCREEN 12:KEY OFF
40   NUMCOLORS=16
50   NUMROWS=200
60   NUMCOLS=200
```

The program will now run in high resolution mode, producing *very* striking pictures of the Mandelbrot set.

It's possible that the BASIC interpreter that you are using doesn't know which graphics adapter you have in your machine. If this happens, you will receive an error message when you try to run this program. The only solution is to use different graphics modes, until you find one that works.

If the numbers in the lower half of the screen start to display in exponential notation (such as 1.342343E-9), it means that you've zoomed in to a level of detail in the Mandelbrot set that this program can't handle. A simple modification to the program will help:

```
10  DEFDBL A-Z
```

However, this will also slow the program a little; this is why we haven't done it. You can make this change yourself should it become necessary.

When the program is first run, it prompts you to type in coordinates for the lower left-hand corner of the image. These consist of the real and imaginary parts of the number that will be displayed as the dot in that corner of your screen. Then it will prompt you for the width of one side of the image. This is the range of the Mandelbrot set that will be visible in the image. The first time that we run the program, let's give it a real coordinate of -2, an imaginary coordinate of -2, and a width of 4. That will display the entire Mandelbrot set in one image.

The program isn't terribly fast. But it's written so that the image will only fill one-fourth of the computer's screen, to save time drawing the image. And the higher the resolution it

Figure 7.2.
Result of running Mandelbrot program with coordinates (-2, -2, 4). This reveals the complete Mandelbrot set.

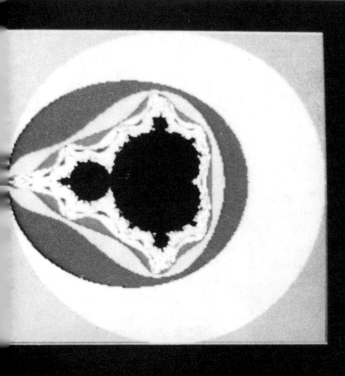

Percentage complete = 100
Dwell for last pixel = 1
Largest dwell = 148
MaxDwell= 150
Real part = -2
Imaginary part = -2
Side =   4

draws the picture in, the longer it will take to draw it. However, the first image of the Mandelbrot set should graph fairly quickly (though you may want to get up and take care of some minor chore while it's being drawn). The numbers at the bottom of the screen tell you how the drawing is proceeding. The "Percentage Done" figure let's you know how much of it has been drawn. The number called the "dwell" is used by the program to determine the color of the dots on the screen. We'll have more to say about the dwell in a moment.

When finished, the graph should look like Figure 7.2. This is the famous insectlike image of the Mandelbrot set. It looks roughly like a ball connected to a black heart. The ball is sometimes referred to as the "head" and the heart as the "body"; the cleft at the end of the heart is called the "inflection point." Everything colored black in this picture (or blue, if you are using the four-color CGA version) is part of the Mandelbrot set. All other colors are in the surrounding areas. Most of the colorful detail is very close to the set itself. In fact, it is the area at the rim of the Mandelbrot set that is the most fascinating part to explore, because it has the most detail packed into it. You can't see that detail at this level of magnification, but as you zoom in deeper and deeper, it will appear. (Unfortunately, as you zoom in deeper and deeper, drawing the pictures will also take longer, for technical reasons that we won't go into here. Eventually you'll reach a point of diminishing returns, where it takes so long to draw the picture that it isn't worth the effort. Check the section For Further Reading at the end of this book for a suggested method of speeding up the process.)

Notice how there appear to be tiny black objects growing out of the side of both the head and the body? These are tiny images of the Mandelbrot set itself—"Mandelbuds," as they are sometimes called. Let's zoom in on one of the Mandelbuds. Press any key and the program will stop. Run it again and zoom in on the area (− .552, .517, .8): that is, tell it you want a real coordinate of − .552, an imaginary coordinate of .517, and a width of .8.

The Mandelbud should look like the image in Figure 7.3. (Actually, Figure 7.3 should look a little more detailed than the image you see on your screen, because it was made using a version of the program that produces full-screen images of the Mandelbrot set rather than squashing those images into a corner of the picture.) Notice how the Mandelbud looks *almost*, but not quite, like a miniature picture of the Mandelbrot set itself. If you're using the multicolor version of the program, look at the bright tongues of color shooting off from the head of the Mandelbud. Run the program again and zoom in on one of these tongues, at $(-.161, .884, .14)$.

This picture, which you can see in Figure 7.4, is even more spectacular than the last. On a color monitor, you can see a tongue of color directly above the Mandelbud's head. See that dark object in the middle of it? It's a mini-Mandelbrot, a tiny image of the Mandelbrot set floating in a sea of color. (Mathematicians interested in the Mandelbrot set believe that all of these mini-Mandelbrots are connected by tiny filaments of the Mandelbrot set that are too thin to be seen at all but the highest levels of magnification. Since our program can't generate magnifications that high, there is no way that we can see the filament connecting this mini-Mandelbrot with the Mandelbud below it.)

Zoom in on the mini-Mandelbrot at $(-.114, .917, .017)$. You'll see it surrounded by what seem to be tongues of flame leaping to all sides. (See Figure 7.5.) It looks almost like a picture of an exploding star in a science fiction novel.

You can zoom in and explore more and more detail in this picture by changing the numbers that you type when the program is first run. Here's the trick: to move the scene left and right, make the first number (the real coordinate) smaller and larger, respectively. (Remember that numbers on the y axis increase upward and decrease as you go down.) To move it up and down, type larger and smaller numbers, respectively. (Numbers on the x axis increase as you go right and decrease as you go left.)

How much should you change these numbers? That's the

trickiest part. The third number represents the width of the picture you are looking at. It tells you how wide the picture is: that is, how much of the complex plane it covers. To move halfway across the picture in any direction, change the appropriate coordinate by half the current width. To move one-third of the way across the picture, change it by one-third the current width. And so on. For instance, if the width is currently .2 and you want to move halfway across the picture to the right, then increase the real coordinate (which moves you left and right) by .1, which is half of .2.

To shrink the window and zoom in on smaller and smaller details, just reduce the width. If the current width is .3 and you want a window one-sixth as large, change the width to .05, which is one-sixth of .3. By changing both the coordinates and the width, you can zoom in on any part of a picture.

This isn't as hard as it sounds. Experiment with the program until you get the hang of it. Note that real and imaginary coordinates outside the range -2 to $+2$ won't produce very interesting images.

Another problem with zooming into deeper and deeper images is that you'll need to change the program slightly in order to see all the details. See those numbers on the screen called Largest Dwell and MaxDwell? The second of these, MaxDwell, is a variable within the program. It determines how much work the program puts into deciding whether a given point in the complex plane is a member of the Mandelbrot set or not. We've set this number initially to 150, which is fairly low. This means that some points that are not in the Mandelbrot set slip through and are painted black on the screen, as though they *were* in the Mandelbrot set. And this, in turn, means that you see less colorful detail when you use this program.

Why did we set this number low? Because setting it to a

Figure 7.3.
Enlargement of Mandelbud

Figure 7.5. (Above)
Exploding star?
Figure 7.4. (Left)
"Tongue" of color above head of Mandelbud

higher number slows the program—and the program already runs slowly enough. When you get down to lower levels of detail, however—especially if you start seeing large, shapeless black blobs on the screen that don't look like part of the Mandelbrot set—you'll need to raise this number. Another clue that you need to raise it is when the number called Largest Dwell on the screen becomes equal to the number called MaxDwell. If both of these numbers are 150 (or if one is 149 and the other 150), then you are probably missing a lot of detail in the picture.

At this point, you should list the program and change the value of the variable MAXDWELL in line 30 to a higher number and you should start to see more detail. What number should you change it to? That depends on how deeply zoomed you are. At first raise it by 100 or 200. Eventually you may need to raise it to at least 1,000 and perhaps much higher. (*Note:* When typing numbers greater than 999 in a computer program, *do not* include commas in the number. For instance, you would type the number 1,000 as 1000.) Raising this number will slow the program, but it's necessary if you want to see all the details of the Mandelbrot set.

Now let's zoom in on another area of the set. Run the program and look at area (− .975, .045, .4) as in Figure 7.6. This is a close-up of the "neck," the area between the head and the body of the set. It is sometimes called sea horse valley, because there are shapes within it that resemble sea horses. You have to zoom quite a way in before you see them, though. Typing coordinates (− .761, .26, .07) will show you a detail of the features along one "cliffside" bordering the valley. (See Figure 7.7.) (Increase the value of MAXDWELL if you begin to lose detail when you zoom.)

Figures 7.8 and 7.9 show other details of the Mandelbrot

Figure 7.6.
Inside sea horse valley

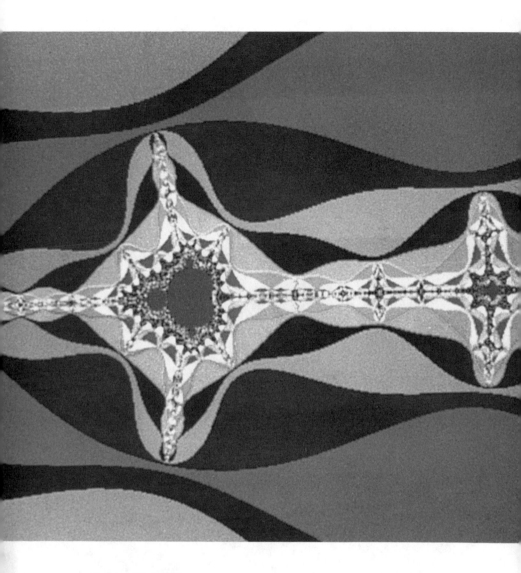

Figure 7.7. (Left)
Detail of sea-horse valley cliff
Figure 7.8. (Above)
Detail of filament in far left part of
Mandelbrot set.
Coordinates are (−1.8902, −0.1368, .37).

set. By running the program and typing in the numbers under these pictures, you can see all of these images on your own computer. And feel free to explore on your own.

What does the Mandelbrot set have to do with chaos theory? For one thing, it is chaotic. It is impossible to predict whether a point on the complex plane will be in the Mandelbrot set without actually performing the proper mathematical operations on that point. There is no shortcut to determining the shape of the set except to graph it one point at a time.

Yet within its seemingly chaotic randomness, it displays remarkable forms and patterns. And new forms and patterns appear at whatever level we zoom into. These new forms and patterns resemble the forms and patterns at earlier levels, but they are never quite the same. They never repeat. And you can keep on zooming infinitely, at least in theory. (All computer programs, including the ones in this book, have a limit as to how far they can zoom in practice.)

Looking at the Mandelbrot set is like looking at pure chaos: unpredictable yet orderly, strange yet familiar. It is a kind of planet Chaos that you can explore to your heart's content.

Figure 7.9.
Another detail of the filament.
Coordinates are (−1.6063, −0.0081, .022).

Epilogue

What does chaos theory mean to modern science? Is it a true revolution in the way we see the universe? Or just a useful tool for exploring systems formerly thought to be unpredictable and unexplorable?

Perhaps it's a little of both. Chaos theorists have so far uncovered no new fundamental laws governing the way the universe works as physicists did earlier in this century, when quantum mechanics was introduced. What they have done, instead, is to give scientists in a wide variety of fields a way to step back from the details of their data and look at the whole thing in new ways, to see new patterns in their data that nobody ever noticed before.

More than anything else, chaos is a new point of view, a new angle from which scientists can view familiar phenomena. In the end, this new viewpoint may provide fundamental new insights into the working of the universe. Or it may simply prove to be a better method of arriving at old insights.

Whichever it turns out to be, chaos theory would seem to mark an important turning point in the history of science. Just how important, only the coming decades will tell.

Typing BASIC
Programs

BASIC (short for Beginner's All-Purpose Symbolic Instruction Code) is a computer language supplied free with most microcomputers. All but a very few of the programs in this book require that you use the version of BASIC supplied with the IBM-PC and compatible computers, so that's the version we'll concentrate on in this appendix.

Before you can type a BASIC program, you must first run a computer program called a BASIC interpreter. On the IBM PC, the BASIC interpreter is called, simply, BASIC or BASICA. On PC-compatibles, it's called GWBASIC. To run the BASIC interpreter you should turn on the computer and wait for the following prompt to appear on the screen:

    C>

or something very similar. (If you do not have a hard disk drive in your computer, you'll see the same prompt with the letter *A* in place of the letter *C*. And you'll need to insert a special disk in the drive with the BASIC interpreter on it.) Now you can type the following:

    GWBASIC <ENTER>

where <ENTER> means that you should press the key on the keyboard marked ENTER (or RETURN). If this doesn't work, try typing BASIC and BASICA. If neither of these works, you may need to ask someone who knows about computers how to run the BASIC interpreter.

There's a second version of IBM BASIC known as QBASIC that is supplied with the most recent version of the computer and is available with some older computers too. If you have a copy of QBASIC, you should use it in preference to GWBASIC, because it makes typing programs easier and runs the programs at a somewhat faster speed. To run QBASIC, simply type:

```
QB <ENTER>
```

(Some versions simply require that you type QB <ENTER>.) Once you've run the BASIC interpreter, you can type the programs exactly as listed in the book. If you're using QBASIC, just type them as though you were using a word processor. Use the arrow keys and the backspace key to correct mistakes.

If you're using a BASIC interpreter other than QBASIC, correcting errors is a little more difficult. Type

```
LIST <ENTER>
```

to see your program and locate errors in it. Type

```
LIST XXX-XXX <ENTER>
```

when you want to see only certain lines of your program, where XXX and XXX are the numbers of the first and last lines that you want to see. (The line numbers are the numbers at the head of every program line you type.) Use the cursor arrows to move to the line and correct any mistakes. Be sure to press the ENTER key while the cursor is still over the line that you've corrected, or the BASIC interpreter will not save your correction.

To delete a line, simply type the line number and press the ENTER key. Or type:

```
DELETE XXX-XXX
```

where XXX and XXX are the first and last lines that you want to delete.

When you've finished typing and correcting your program, move the cursor to a blank line and type

    RUN <ENTER>

This will cause the program to be executed. If you want to stop the program prematurely, press the key marked BREAK while also holding down the key marked CONTROL. The program will stop. (Sometimes you may have to press ENTER before you are returned to the BASIC interpreter.)

And that's all there is to it. If you'd like to know more about writing BASIC programs of your own, there should be a number of books in your library and in your bookstore telling you how to do it—including several written by the author of this book.

Source Notes

1. R. A. Kerr, "Does Chaos Permeate the Solar System?" *Science,* April 14, 1989, 144.

2. R. A. Kerr, "Pluto's Orbital Motion Looks Chaotic." *Science,* May 20, 1988, 986.

3. R. A. Kerr, "Does Chaos Permeate the Solar System?"

4. R. Pool, "Is It Chaos, or Is It Just Noise?" *Science,* January 6, 1989, 25.

5. R. Pool, "Is It Healthy to Be Chaotic?" *Science,* February 3, 1989, 604.

6. K. McAuliffe, "Get Smart: Controlling Chaos," *Omni,* February 1990, 42.

7. W. J. Freeman, "The Physiology of Perception," *Scientific American,* February 1991, 78.

8. R. Pool, "Is It Healthy to Be Chaotic?"

9. R. Pool, "Is Something Strange about the Weather?" *Science,* March 10, 1989, 1290.

Glossary

Attractor—When an iterated equation (see definition) or set of iterated equations is repeated a certain number of times, it may settle down on a single value that is repeated over and over; this value is called an attractor. Some equations have multiple attractors, which are repeated over and over again in a pattern.

Bifurcation—The transition of an iterated equation from a single attractor to more than one attractor.

Bifurcation diagram—A diagram that visually depicts the transition of an iterated equation from a single attractor to more than one attractor.

Butterfly effect—The half-joking term used by meteorologists and other scientists involved in predicting complex systems to describe the large effect that can sometimes be produced by small causes, such as the theoretical effect of a butterfly flapping its wings in Asia on the weather in New York some days later.

Chaos—The term used by scientists to describe a system in which small changes in initial conditions can have a significant and unpredictable effect on the eventual outcome.

Complex numbers—Numbers that are expressed as the sum of two numbers, one called the real part and the other the imaginary part. For example: 7 + 19*i*.

Complex plane—The two-dimensional area in which all complex numbers can be graphed.

Convection—The way in which heat moves through air, and certain other substances, by creating motion among heated molecules.

Deterministic—Based on the certain, rather than the probable, outcome of events.

Differential equations—A type of equation used to describe objects in constantly changing motion.

Feigenbaum's number—A ratio, expressed as the number 4.669202609, and discovered by scientist Mitchell Feigenbaum, that occurs in the bifurcation diagrams produced by many different equations.

Fractal—Short for "fractional dimension"; describes objects that have a number of dimensions, such as 2.7, which cannot be described as a whole number.

Graph—A drawing that demonstrates the changing relationships between numbers in an equation.

Graphics—Pictures on the video display of a computer.

Iterated equation—An equation that must be solved repeatedly, with the value of one of the variables on each iteration (or repetition) becoming the value of one of the other variables on the succeeding iteration.

Logistic difference equation—A common equation used in the sciences and represented as: $N_{T+1} = RN_T (1 - N_T)$

Lorenz's butterfly—The butterfly-shaped graph of the strange attractor produced by Edward Lorenz's equations describing an imaginary weather system.

Mandelbrot set—The set of complex numbers that behave in a certain way when subjected to an iterated equation used by the scientist Benoit Mandelbrot.

Noise—Interference from external phenomena that have nothing to do with a specific phenomenon being studied.

Number line—A line used to visually represent the relationship of numbers to one another, along which traditionally the numbers grow larger to the right and smaller to the left.

Probabilistic—Based on probabilities rather than certainties.

Probability—The percentage chance that an event will occur or not occur.

Program—As a noun, a series of instructions written in a specially designed language and placed in the memory of a computer, which tells the computer what it is supposed to do. As a verb, the act of writing such a program.

Sensitive dependence on initial conditions—Another name for the butterfly effect; the tiny differences in two sets of data don't cancel each other out, and tiny errors can result in big errors.

Simulations—Mathematical descriptions of systems that exist in the real world. These descriptions are commonly used to create computer "models" of these systems.

Strange attractor—An attractor (see definition) that consists of a self-similar pattern of numbers that never repeats exactly.

Three-body problem—A classic problem in physics that involves describing the effect that three bodies have on one another when all are producing a force, such as gravity, that affects the others. The three-body problem is considered nearly impossible to solve precisely, though approximate solutions can be achieved.

Three-dimensional graph—A graph with three axes—an *x* axis, a *y* axis, and a *z* axis—each at a right angle to the others.

Two-dimensional graph—A graph with two axes, an x axis and a y axis, each at a right angle to the other.

Variables—Numbers in an equation that change (or *vary*) according to the conditions being described by the equation.

X axis—The horizontal dimension of a graph, usually represented by a horizontal line with numeric callibrations on it.

Y axis—The vertical dimension of a graph, running at a right angle to the x axis and any other axes in the graph.

For Further Reading

Briggs, James and F. David Peat. *Turbulent Mirror*. New York: Harper & Row, 1989. An interesting if rather mystical introduction to chaos theory by a pair of physicists. Profusely illustrated and contains some interesting descriptions of chaotic phenomena, but don't take the mystical trappings too seriously.

Gleick, James. *Chaos: Making a New Science*. New York: Viking, 1987. This is a nontechnical but still rather difficult book on chaos theory written for a popular audience. It was a surprise bestseller when it was published in 1987 and is worth checking out of the library if only for the full-color pictures of the Mandelbrot set in the photo insert.

Sheffield, Charles. "The Unlicked Bear Whelp," in *New Destinies*. New York: Ace Books, 1990. This article, which appeared in a paperback science fiction magazine, contains more solid information about chaos theory than some entire books written on the subject (though I won't name any names). Though Sheffield gets a little technical, there are enough interesting insights here to make this worth digging out at your local library or bookstore.

Tyler, Bert, et al. *Fractint*. The Stone Soup Group, 1990 and later. *Fractint* is a computer program distributed free for the IBM-PC and compatible computers by a collective of programmers calling themselves the Stone Soup Group. It can be obtained at most computer user's groups or by downloading from computer bulletin boards and on-line services such as CompuServe and GEnie. It produces beautiful drawings of the Mandelbrot set, bifurcation diagrams, and lots of other fractal images—and it draws them much faster than the programs in this book. Well worth getting a copy; the price (free) is certainly right.

Index

Animal populations. *See* Populations

Asteroids, 74–75

Astronomy, 59–61, 74–76

Atoms, 24

Attractors, 44–46, 50. *See also* Strange attractors

BASIC computer languages, 27, 111–113

Bifurcated systems, 46

Bifurcation diagrams, 48–57, 87

Body, 78–80, 85

Brain waves, 79–80

Butterfly effect, 68–69

Chaos, 13–14
  in astronomy, 74–76
  in epidemiology, 76–77
  and fractional dimensions, 72–73, 85
  in human body, 78–80
  and Mandelbrot set, 107
  in population equations, 41–42, 45–55
  significance of, 109
  in stock market, 82–83
  in transcendental functions, 56–57
  in turbulence, 24, 82
  in weather, 65–68, 80–82

Charm, 17

Coastlines as fractals, 85

Complexity, 21, 24–25, 58

Complex numbers, 88–89
Complex plane, 88–90
Computers, 16, 25. *See also* Programs, computer
  and predictions, 61–68
  programming of, 20–21, 26–31
Consciousness, 79–80
Contagious diseases, 76–77
Convection, 69–71
Creative thought, 79

DELETE computer command, 112
Deterministic predictions, 64–65
Differential equations, 59–60
Digital orrery, 75
Dimensions, fractional, 72–73, 79. *See also* Fractals
Diseases, 76–77
Distance formula, 18–19
  graphing of, 31–36
  program for, 27–31
Dwell, 96, 99, 103

Earth, orbit of, 76
Economics, 84–85
Electrocardiograms, 78
Epidemiology, 76–77
Equations, 17–19
  and complexity, 58
  differential, 59–60
  for diseases, 77
  generality of, 23–24

graphs of, 30–36
iterated, 39–40
logistic difference (*See* Logistic difference equation)
and noise, 24–25
in programs, 20–21, 26–27
and strange attractors, 72
Errors, 60–61, 63, 67–69

Feigenbaum, Mitchell, 58
Feigenbaum's number, 58
Fractals, 84–87, 89–90
Fractional dimensions, 72–73, 79
Fractional numbers, 89
Freeman, Walter, 79

Goldberger, Ary, 78–79
Graphs and graphics, 30–36
  of convection, 69–71
  of Mandelbrot set, 88–108
  of populations, 37–38, 47–55
Gravity, 59–61
Great Red Spot, 80–82

Heart, human, 78
Human body, 78–80, 85
Hyperion, orbit of, 75

Imaginary numbers, 88
INPUT computer command, 28

Inspirations, 79
Iterated equations, 39–40

Jupiter, 74, 80–82

La Place, Pierre Simon de,
61
Laskar, J., 76
LIST computer command,
112
Logistic difference equation
and Feigenbaum's
number, 58
graph of, 47–55
for populations, 39–41
program for, 42–46
for weather, 69–71
Lorenz, Edward, 65–71
Lorenz's butterfly, 71

Malthus, Thomas, 37
Mandelbrot, Benoit, 84–90
Mandelbrot set, 16, 87–108
Mandelbuds, 96–97
Marcus, Philip, 80, 82
May, Robert, 37–40, 55
Meteorites, 74–75
Meteorology, 64–68,
80–82
Mini-Mandelbrots, 97
Models, computer, 21,
26–27
of populations, 40–41
of solar system, 75
of weather, 65–68
Moons, motion of, 23, 59–
61, 74–76

Newton, Isaac, 23–24, 59
Noise, 24–25, 118
and chaos, 46–47
and populations,
40–41
Number lines, 88
Numbers
and computers, 26
and equations, 17–18
on graphs, 30–32
in Mandelbrot set,
87–90

Orbits, 59–61, 75–76
Output, computer, 27

"Period Three Implies
Chaos" (May), 55
Pictures, computer. See
Graphs and graphics
Planes, 72, 88–90
Planets, 59–61, 74–76
Pluto, 75–76
Populations
equation for, 39–41
graphs of, 37–38,
47–55
program for, 42–46
Predictions, 12–16
and computers, 61–64
with Mandelbrot set,
107
of population, 42–46
of solar system, 59–
61, 74–76
of weather, 64–68,
80–82

PRINT computer command, 28
Probabilistic predictions, 64
Programs, computer, 20–21, 26
  for distance formula, 27–31
  for graphs, 31–36
  for logistic difference equation, 42–46
  for Mandelbrot set, 91–108
  for transcendental functions, 56–57
  typing in, 111–113

QBASIC language, 112

Rapp, Paul, 79
Real numbers, 88
Repetitive computer calculations, 29–30
Resolution in programs, 93
Rounding errors, 63, 67–69
RUN computer command, 28, 113

Saturn, 75
Schaffer, William, 77
Self-similar patterns, 84–85
Sensitive dependence on initial conditions, 69
Simplicity, 21, 24
Simulations, 21
SIN function, 56–57

Solar system, 59–61, 74–76
Stock market, 82–83
Strange attractors, 71
  and fractional dimensions, 72–73, 85
  and stock market, 82–83
  and weather, 82
Strangeness, 17
Subscripts, 39
Sussman, Gerald, 75–76

Thoughts from chaos, 79
Three-body problem, 60
Three-dimensional graphs, 71
Tick marks, 34–35
Transcendental functions, 56–57
Tsonis, Anastasios, 82
Turbulence, 24, 82
Two-dimensional graphs, 32

Variables, 18–19, 27, 32

Weather, 64–71, 80–82
Wisdom, Jack, 74–76

X axes, 30–35, 89

Y axes, 31–35, 89
Yorke, Jim, 55

Zooming in programs, 52, 96–99, 103

About the Author

Noted science writer Christopher Lampton has written more than fifty books on such subjects as astronomy, computers, genetic engineering, and the environment. His most recent books for Franklin Watts are *DNA Fingerprinting, Telecommunications: From Telegraphs to Modems,* and *Gene Technology.* In addition to his nonfiction, he has written four science fiction novels. Christopher Lampton holds a degree in broadcast journalism and makes his home in Silver Spring, Maryland.